SAVING MORE THAN SEEDS

T0265051

Saving More Than Seeds

Practices and Politics of Seed Saving

CATHERINE PHILLIPS

Routledge
Taylor & Francis Group

LONDON AND NEW YORK

First published 2013 by Ashgate Publishing

Published 2016 by Routledge
2 Park Square, Milton Park, Abingdon, Oxfordshire OX14 4RN
711 Third Avenue, New York, NY 10017, USA

First issued in paperback 2016

Routledge is an imprint of the Taylor & Francis Group, an informa business

British Library Cataloguing in Publication Data
A catalogue record for this book is available from the British Library

The Library of Congress has cataloged the printed edition as follows:
Phillips, Catherine, 1973–
  Saving more than seeds: practices and politics of seed saving / by Catherine Phillips.
    pages cm
  Includes bibliographical references and index.
  ISBN 978-1-4094-4651-4 (hardback: alk. paper)
  1. Seeds—Storage. 2. Seeds—Harvesting. I. Title.
  SB118.4.P45 2013
  631.5'21—dc23
                                                                2013007753

ISBN 13: 978-1-138-27180-7 (pbk)
ISBN 13: 978-1-4094-4651-4 (hbk)

Contents

List of Figures

List of Tables

List of Acronyms

CBD	Convention on Biological Diversity
CFIA	Canadian Food Inspection Agency
CSGA	Canadian Seed Growers' Association
CSTA	Canadian Seed Trade Association
FAO	Food and Agriculture Organization of the United Nations
GCDT	Global Crop Diversity Trust, also 'the Trust'
GE	Genetically Engineered or Genetically Modified
IPR	Intellectual Property Rights
ITPGRFA	International Treaty on Plant Genetic Resources for Food and Agriculture, also 'the Treaty'
LLP	Low-level Presence
PBR	Plant Breeders' Rights
PGR	Plant Genetic Resources
PGRC	Plant Gene Resources of Canada
PGRFA	Plant Genetic Resources for Food and Agriculture
SGSV	Svalbard Global Seed Vault, also 'the Vault'
TRIPs	Trade-Related Aspects of Intellectual Property Rights
UPOV	Internationale pour la Protectiondes Obtentions Végétales [International Union for the Protection of Plant Varieties]
US	United States of America

Acknowledgements

I am indebted to many for their support and assistance in various stages of this project. I met many passionate, helpful people in the course of researching and writing this book and it is impossible to thank them all individually. But I am grateful for each encounter. Special thanks go to those savers who shared their thoughts and practices, welcoming me into their saving lives and spaces. Without them, this work would not have been possible.

I have benefitted also from ongoing conversations and particular engagements with academic colleagues. For those of you who took the time to discuss ideas with me in conferences and classrooms, corridors and cafes, and over computer lines – thank you. I have been fortunate to work with many generous and insightful people in the Australian Centre for Cultural Environmental Research at the University of Wollongong (Australia) and in the Faculty of Environmental Studies at York University (Canada). Particular thanks go to Liette Gilbert and Lesley Head – each of whom supervised my work during different parts of this project. For productive discussions and comments I also thank: Elisabeth Abergel; Jenny Atchison; Chantel Carr; Jen Cypher; Leesa Fawcett; Leah Gibbs; Chris Gibson; Eric Higgs; Brewster Kneen; Cathleen Kneen; Mustafa Koc; Rod MacRae; Megan Salhus; Catriona Sandilands; Leela Viswanathan; and Gordon Waitt. For our work together on seed policy, I am grateful to Brewster Kneen, Devlin Kuyek and Sarah Martin.

Thanks also to seed saving colleagues for enriching my experiences through workshops, conferences, workdays, seed fairs and such. From this diverse group of people and organisations, my gratitude goes especially to savers at the Heirloom Seed Sanctuary, Bob Wildfong at Seeds of Diversity Canada, and the staff of USC Canada, particularly Kate Green, Lise Latremouille, Jane Rabinowicz, Awegetchew Teshome and Susie Walsh. Thanks also to the many participants who came from around the world to share insights about seed sovereignty at the *From Seeds of Survival to Seeds of Resilience* conference in Ethiopia.

I am also grateful for the financial support this project has received. This project was partially funded by a grant from the Australian Research Council to Lesley Head (FL0992397). In addition, throughout my dissertation studies supplementary funds came through several grants and awards from York University and its Faculty of Environmental Studies.

For generously agreeing to have their photo or graphic published in this book, my appreciation goes to Jen Cypher (Figure 9.1), Phil Howard (Figure 3.1), Christina Lowry (cover photo), and Bob Wildfong (Figure 7.3).

Finally, I thank Anne and Len whose inspiration and love keep me going; my extended family – especially Lynn, Mary, Gord, Brett and Leanne – for their unfailing support even in times when their difficulties were far greater; Jen, Laine and Owen for putting up with me, and putting me up; and lastly, Scott for making my world better every day. Thank you.

Earlier iterations of some of the research presented here can be found in:

Phillips, C. 2005. Cultivating practices: saving seed as green citizenship? *Environments*, 33(3), 37–49.

Phillips, C. 2008. Canada's evolving seed regime: Relations of industry, state and seed savers. *Environments*, 36(1), 5–18.

Chapter 1
Starting with Seed

In the simple act of planting I was engaged in one of the most universal – and certainly one of the most important – of all human activities. I share the act of planting and my hope for a harvest with most of the world's population and with unnumbered previous generations. People must eat. And the chain of production processes that finally delivers food to our mouths – long for the New Yorker, short for the Thai peasant – begins everywhere with the sowing of the seed.

Jack R. Kloppenburg Jr., *First the Seed*

When I was a child I engaged in an experiment with my class, an experiment common in grade schools across Canada (perhaps other places too). In the exercise, we took several bean seeds and pressed them against the sides of a glass jar with a moist paper towel. The glass jar was placed in a sunny window. We made sure to keep the paper towel moist, and were rewarded as the miracle of sprouting was slowly and amazingly revealed. I remember that some of the children were surprised and delighted by the idea that plants (and ultimately food) came from seeds. This insight was not news to me.

My father, continuing the subsistence practices of his parents, had a food garden while I was growing up. For us the produce was not necessary for survival, but caring for the crops was taken seriously (though we often had fun doing it). I watched my father hand-sow his garden and another at my grandmother's place and, when I was not busy making mud pies or playing with rabbits, bugs or other children, I helped sow the seed and care for the plants. My father sometimes even went to the trouble of marking the area of seeds I had planted so that I would be able to see what resulted. Sometimes I enjoyed gardening, other times it felt like a chore. As an adult though, I remember it fondly. So, it was not news to me that seeds grew into plants that made food. However, while I watched and tended my beans in the school window, it was quite wonderful to see the process that had been hidden from my view by soil.

The growth of roots, the splitting of the seed, the quest of the sprout for light was revealed day by day and I watched very intently. The lesson, of course, was meant to relay the practical use of plants in supplying our food and to convey some of the biology of plants, and it achieved those aims. But, for me, the experiment also brought with it an engagement with the amazing abilities of seeds.

* * * Flash forward twenty-odd years * * *

You find me again in a classroom, doing an exercise. But now I am the one facilitating, in a university class. Here is the task I outline: Imagine the last 'meal' you had. Draw it. (I know, lots of people get nervous at this point. Do it anyway. It is an exercise not a test. I'm doing it too, and I am not a skilled artist.) Now, somewhere around that drawing list the things that went into that meal. Keep it simple. If the last thing you consumed was coffee and toast, for instance, you might write: coffee beans, water, sugar, milk, bread, butter. If you are feeling ambitious, you might break it down further – bread into flour, water, yeast, salt, sugar and butter, for instance – or you might add things that were part of the process – electricity, cup, knife and so on. But really, it will get complicated soon enough, so you might want to save that for later.

Take some time to think about the ingredients you have listed. Try to follow them back through their journeys to you – through retailers, distributors, producers and such. Put those on your drawing somehow. Where will the things go after you? Add that too. Now ask yourself some questions. Are there gaps in your knowledge? Where? Is there anything you are surprised that you don't know, or that you do? Anything you would like to find out? Who produced those fair trade coffee beans perhaps? What standards they had to meet? Where do all the grounds go? What might they be used for? Do you know some of coffee's history – like where it was domesticated? Or perhaps how diverse the crop is? And what about the water? You get the idea. Talk to your neighbour, think it through together.

I have used a variation of this exercise in several social sciences classes now – some oriented specifically around food, others with more distant connections. Sometimes a short version provides a start or break for a lecture, other times it is part of deeper investigations done over the year. The exercise points to one of the approaches often used these days to understanding food – looking at a particular item from 'field to table' or as part of a commodity chain. It is a version of 'following-the-thing' in which a commodity's paths, altered forms, revaluations and connections are traced (Cook et al. 2006). This exercise, in its various forms, has evoked some fascinating conversations, and some inspiring projects. But often seeds are hiding in the resulting stories. In that listing of coffee beans, for instance, it could just as easily have read coffee seed. It would in one sense be more accurate – since those beans are really seeds from fruits of the coffee tree that have been dried and roasted. But seeds transform into plants or become food, and sometimes their lives as seeds are forgotten along the way.

The combined utility and wonder of seeds was highlighted for me in that early school experiment, but often this importance of seeds is ignored as an obvious, everyday thing or obscured as part of distant agricultural systems. Kloppenburg (2004) at the opening of this chapter reminds us of the importance of seed. The transformations of wild seeds into cultivated ones that are saved and resown year after year fundamentally altered both plants and people. Domestication goes both ways, in complex interrelations among seeds and savers.[1] Humanity's survival – biological and cultural – has become tied up with the lives of seeds. Seeds are indispensable as a means of reproducing food, as food themselves, as part of ecosystems that support and constrain us, and as part of our cultural heritages. Seeds (and their plants) are part of the socionatural challenges we face in loss of biodiversity, maintaining food security, adapting to climate change, and sustaining rural and urban livelihoods. The histories and destinies of both seeds and people have become entwined. Questions of how people relate with seeds are vital to understanding what agrifood possibilities arise or are shut down.

Food production begins with seed sowing, as Kloppenburg (2004) and my classroom exercises suggest, but this does not tell the whole story. The act of planting seed to get food (or gardens) seems simple. Seeds are planted, they grow into plants, plants (or parts of them) are eaten, plants produce seeds, seeds are stored, repeat. But the seeds being sowed, the knowledge of how to grow and save them, the methods and tools used, and uses of seeds are not at their beginnings. Rather, they are always changing and exist always in relation with others. Savers choose seeds for things like textures and tastes, adaptability and appearance, storability and cultural suitability, making changes to ecologies and agrifood cultures as they do. And seeds, for their part, alter human physiologies, tastes, cultures and production practices.

Seeds, like people and other nonhumans, have histories that condition their existences and their present embodiments suggest their possible futures. Visible or not, seeds exist as part of plants, as potential foods, as governed beings, as a metaphor for beginning anew, and more besides. How seeds are spoken about and viewed depends on the time in their lifecycle as well as such factors as the experiences, purposes and knowledges of the persons involved. Even at the same point in a lifecycle, seeds can be renamed based on their purpose – dried beans or poppy seeds ready for cooking, for example, are also seeds waiting to be planted. Shared labours of seeds and savers, bringing together their relative capacities and dispositions, results in the diversity and interdependencies at the base of today's food and agriculture.

Seed saving provides the focus of this book. I use 'seed saving' as a shorthand term for a complex set of practices including the planting, tending, harvesting, storing, eating and replanting of seeds (and other propagating material),[2] as well as the attendant processes of exchanging and knowledge-building. 'Seed savers',

---

1   For reflections on domestication as interrelation see Anderson (1997); Head and Atchison (2012: ch.3); Tsing (2012).

2   The 'seeds' of seed saving include those organisms so defined biologically, but also a broader category of propagators. Most saving is done with seeds – small embryonic plants with stored food in a coat. But these are joined by things like seed potatoes, garlic cloves, rosemary cuttings, ginger rhizomes and so on. My use throughout the book defers to how savers generally use the term. There are times of confusion and of distinction, which I try to note.

Fig. 1.1   Saved seed – poppy and bean

or the men and women growers – farmers or gardeners – around the world who cultivate and keep seeds as primary or ancillary activities repeat these practices year after year. However, seed saving is more than a simple technique. What seems like an obvious and straightforward activity that takes place in various gardens and farms around the world is, like the seed itself, far more significant and complicated than it may appear. The art and science of seed saving are fundamental to our agrifood networks, and all of the collaborations of previous savers and seed generations come together in each enactment as savers and seeds reconstitute the cycle anew, looping forward together.

Seed saving is a set of practices that raises vital ethico-political questions: who has access to seed and for what purposes?; whose knowledge is valued and how?; who/what participates in arranging seed relations and how? These questions surrounding seed saving are being asked within international, national and local arenas. They are expressed in louder and more urgent tones as seed saving becomes increasingly constrained and controversial with the reordering of agrifood through, for examples, corporate priorities, governmental policies and changing agricultural practices. Seed saving, once common practice but now often considered outdated, has become an ecological, economic, cultural, political and survival issue for many. I came to do this research on seed saving within this context, and it is why I continue to pose questions about the relations among seeds, growers, eaters, seed saving and agrifood orders.

In this book I follow seed saving through fields, governmental policies, international economies, and elsewhere. I am interested in how seed saving is enacted, enrolled and remade in various ways. The seeds themselves are present in different ways – as commodities, food, biodiversity, living beings and so forth – as are savers – showing up as consumers, activists, conservers and rights holders. In attending seed saving and its practitioners, I explore the idea that seed saving practices may offer ways of living that are vitally different from those presented through neoliberal, corporate orderings. What people and seed do together, how they enact their shared worlds, the meanings that are recreated, and the possibilities that emerge through the process of saving seed form the core of this study.

Thinking Through Seed Saving

Social studies research on seed saving has concentrated on how contemporary restructuring of seed systems toward privatisation, industrialisation and commodification impacts primarily global South countries and peoples in various (usually negative) ways (see Brush 2000, 2004; Fowler and Mooney 1990; Kloppenburg 2004; Shiva 1993, 1997, 2000). In comparison, global North growers, as participants in 'modern' agrifood orders, are expected to stop saving seed and purchase them instead each year as inputs. Studies of seed saving in the global North have begun to trouble binary thinking of this kind (see Mascarenhas and Busch 2006; Nazarea 2005; Vellvé 1992).

While seed saving may be a less popular practice with global North growers than it once was, it is a set of practices with which some farmers and gardeners still engage and that many support. Seed saving is gaining some attention in the global North as savers share their own stories of living with seeds (Jason 2011; Whealy 2011), and academics begin to notice (Carolan 2011: ch.4; Nazarea 2005: ch.4; Phillips 2005, 2008). Hence, studies of seed saving are relevant to the global North as well, in ways both related to, and distinct from, those applicable in global South contexts. Sustained attention to the practices of seed saving, with its implications for savers, seeds, and their shared worlds, is as yet, underdeveloped. This book contributes to developing an understanding of seed saving, particularly within Canada, and situates this within broader international trends.

A range of academic theorists serve as inspiration to better understand seed savers' practices, and the contexts within which they operate. Though some of this book's chapters rely more strongly on particular theorists and ideas, the various conceptions I explore run through the chapters and each other, overlapping and blending in diverse ways. Not all the theorists whose writing I use agree; rather, each chapter offers support, depth and/or connection to other insights. Three theoretical threads run through the book: everyday practices; nonhuman agency; and ethico-political engagement. Versions of these themes that I find particularly helpful are explored more fully in the next chapter, but here I offer a sketch.

To begin, this book evolved from learning about the everyday practices of seed savers. Scholarship on everyday life suggests that lived experience brings together the inane and exceptional, and that it is through practices that support for, and resistance to, dominant societal relations manifest (de Certeau 1984; Lefebvre 1984). Moreover, as we do what we do we recreate the worlds we inhabit and ourselves, the other people and things we relate with, and the worlds in which we engage, each of which, in their turn, influence our possibilities. In this sense, our worldly engagements are always already part of relational worlds in which change, conflict and collaborations occur in multiple ways. In this book part of my interest is exploring how seed saving shapes and is shaped by its practitioners and world relations.

I speak about seed saving as a set of everyday practices, by which I mean first, that it is an activity that requires engagement in the world with other things – like seeds; and second, that this engagement is repeated, never in exactly the same way but recognisably sustained through time and space. There is significant intellectual interest in everyday practices (performances, enactments, et cetera) across social studies (*cf.* Borgmann 1984; de Certeau 1984; Haraway 2008; Hinchliffe et al. 2005; Leidner 1993; Mol 2002; Schatzki 1996; Shove et al. 2009). Highlighting how practices matter, Ursula Franklin argues that what is being done may be less important than how it is being done: 'One has to keep in mind how much the technology of doing something defines the activity itself, and, by doing so, precludes the emergence of other ways of doing "it", whatever "it" might be' (1999: 9). Following Franklin, the doings, knowings and possibilities of genebanking, for instance, differ from those of seed saving, though both may be (in part) aimed at conserving diversity. In these works, and in this book, there is a focus on practice and on the process of how things are done, as well as an expectation of change through altering materialities, senses, knowledges and relations. However, this book examines not only what people do, but also what they think about what they do. The meanings and values that people find through their engagements with seeds as savers have a strong role in this book.

Acknowledging nonhuman agency is vital to broader debates that unsettle dualisms between nature and culture, human and nonhuman, active and passive. By now, many have recognised that nature is not a realm in and of itself, nor is it constructed or controlled solely by people. Environmental philosophers and ethicists have challenged anthropocentrism through arguments in favour of the intrinsic value of individual and collective natures, recognising that beings and things other than humans have purposes of their own and that these purposes should be recognised, encouraged and facilitated (*cf.* Callicott 1986; Leopold 1949; Plumwood 2002b, 2007). Evidence of nonhuman tool-making, communication, planning and altering behaviours based on different scenarios each contradict notions of humans as the only possessor of active agency. The example of crows demonstrating intelligence by using cars travelling down roads to crack nuts for them is, by now, famous (Attenborough 1998). Challenging human-centredness does not suggest that nonhumans have the same agentic capacities as humans; rather, the aim is to find ways of recognising nonhumans, their lives and relations, for what they are in and of themselves, to acknowledge differing and uneven experiences of affecting and being affected. It is increasingly recognised that, as Haraway (1992: 67) states, 'all of the humans are not "us" however defined' and 'all of the actors are not human.'

The idea that agency is more distributed and relational rather than concentrated in people brings together otherwise disparate studies through an interest in 'relational materialities'. An underlying shared interest exists in understanding how different things and beings interact, and how their relations shape themselves, others, and the worlds in which they exist. People, documents, technologies, birds, whatever else, each has the potential to create change in the world through their relations with others. Scholars have discussed in this way the complex relations of things such as bacteria (Hird 2009) and mushrooms (Tsing 2005, 2012), door closers (Latour 1988) and markets (Callon 1991), buffalo (Lulka 2004) and elephants (Lorimer, J. 2010). A few have even addressed seeds in some fashion (Clark 2002; Katz 2006; van Dooren 2012; Whatmore 2002: ch.5, 6).

Seeds do have their own lives, and in some cases present opportunities to people – think for example of a plant going to seed offering a chance to save seed for next year or for the seed to drop to the ground. But it is not only a concern with nonhumans and their agency that is of concern here. While individual and collective agency is relevant, as Mol (1999: 87) reminds us, considerations of agency also require exploration of how particular beings and things are 'defined, measured, observed, listened to, or otherwise *enacted*.' The point then, is to examine orderings and practices, as well as particular agents' actions (positive and negative).

This book is part of ethico-political engagements in two different ways: first, it explores the ethics and politics of seed saving; and second, it is an effort to challenge our thinking about present realities and future possibilities of seed saving. Concerns about (among other things) food, biodiversity, development, genetic engineering and climate change each relate with plants and seeds at various scales and times.

Important implications arise once we allow for nonhuman agency. Efforts to recognise how nonhumans and humans might work together to improve their shared worlds have been elaborated recently for instance in Bennett's (2010) vibrant matters, Haraway's (2008) response-ability and shared suffering, Latour's (2004a) proposed parliament of things, and Hinchliffe and Whatmore's (2006) convivial, cosmopolitical world. These explorations bring together insights from environmental, philosophical, feminist, sociological and geographical accounts. In general, arguments for creating more ethically responsible and sustainable ways of living with nonhumans mandate two interrelated shifts: respecting differences and learning to be open to others, and possible futures. How to achieve these related shifts comes down to mindfully and regularly engaging with others in our everyday lives while 'taking seriously' those with whom we relate. The argument is that 'ontological and ethical considerations are inseparable' (Wolch and Emel 1998: xi), therefore one of our tasks is to make room for nonhumans and be open to their difference. Within these formulations we can see the two themes already outlined as present in this book: nonhuman agency and practices in the everyday.

Our everyday lives may be ordered, disciplined and dominated in particular ways, and we may support some of these relations through our practices; however, everyday practices are also ways of opening and pursuing possibilities of living otherwise (Borgmann 1984; de Certeau 1984; Haraway 2008; Lefebvre 1984, 2002). It is not only that we are each entangled in various orders, some more stable and powerful than others, but that in conflict and cooperation we recreate our worlds through practices. Because relational networks are practiced, or enacted, they are open to change. We do not have to accept specific relations or networks as the only, or even the preferred option; rather, every relation and ordering is contestable and contested. Other possibilities not only exist, but may offer better ways of living together.

This book, then, takes up the call for researchers to pay 'close and respectful attention to the practical knowledges and vernaculars of everyday sense-making' (Whatmore 2002: 162) in efforts to engage in research that is 'avowedly and unavoidably a form of intervention in the world, opening up rather than pinning down, the possibilities' (Hinchliffe and Whatmore 2006: 134). In dealing specifically with seed saving, as connected with the many agrifood and environmental challenges we face, this book is also an effort to contribute to a 'better understanding of the links between embodied individuals (both human and plant) and the broader landscapes that provision them' (Head and Atchison 2009: 237).

Exploring Seed Saving

For this book I collected information on seed saving practices and orderings using several methods: participant observation, as a saver and event attendee; interviews with savers and experts; analysis of published documents; and a survey of savers across Canada.

As part of participant observation methods, I began saving seeds. Most seed saving is done by growers as they go about their everyday activities. In some cases – such as workdays – groups engage, but much of the time it is individuals cultivating gardens or farms. With this in mind, as part of my exploration I decided to include seed saving in my gardening. I was already growing some vegetables and herbs for home consumption, but seed saving meant doing things differently. It meant my using open-pollinated seeds (which most varieties were, but some were replaced); rearranging garden plans to accommodate larger plants for longer times, and cross-pollination issues; tending and observing plants to ensure good seed – not just good food; finding space and learning methods for processing, drying and storing seeds; and much more. In my gardening I focus on food plants, with some flowers added. Because I move relatively frequently, concentrating on annuals – with some perennial herbs in containers – makes sense to me. For most of the research I lived in urban areas in southern Ontario, with some time spent on the West Coast and in the Eastern Townships of Québec. During the research I had several different gardens including combinations of front, back, and side yards; containers on balconies; and community garden plots, by myself and with others. Practicing seed saving as part of this work was important in gaining a more grounded knowledge of what seed savers do and the choices they face as well as being able to communicate with other savers effectively.

My own saving practices were supplemented by my participation in and observation of many workdays, meetings, seed fairs, and informal chats over the garden fence or kitchen table. I found these experiences invaluable. As an example of the diversity involved in this kind of participation, I have attended many Seedy Saturday events over the years (for this research, but also aside from this project). Seedy Saturdays are like seed fairs; they are events organised by small groups of people in various neighbourhoods and regions at which seeds are swapped and sold, practical workshops run, informative booths and presentations, and so on. At various times, I acted as observer, buyer, advice-giver, browser, panel discussant, co-organiser, non-governmental organisation promoter, swap table volunteer, workshop facilitator, and helper for some small-scale seed grower-retailers. Each of these events, usually held during Canada's winter–spring months (January to April), has its own character and offers insights into the region and its seed savers. The other activities I engaged in were just as varied and interesting.

In addition to participant observation, I interviewed seed savers and involved others (genebankers, breeders, non-governmental advocates). The interviews were done over the phone, email, mail or/and in-person – depending on what worked best for participants. The savers I interviewed called themselves gardeners, farmers and hobby growers. Though there are seed savers within various scales and production systems of agriculture in Canada, the interviewees for this book were small-scale growers; the largest area under cultivation was estimated at three acres (though the properties were often larger). Saver interviewees varied in their practices: some sold seed, some did not; some had been saving for their whole lives, some only for a short time; some saved a few varieties, others hundreds (see Appendix I for characterisation of interviewees). These interviews occurred as a series, beginning with semi-structured questions and progressing to more individualised follow-up questions. After doing a preliminary analysis, all the saver interviewees were sent a set of themed interview quotes and given the opportunity to comment.

Throughout this book, I quote seed savers who shared their thoughts and practices with me. These quotations are meant to be indicative of the ways in which seed saving is understood and practiced. However, in using categories of actors – seed savers, growers, genebankers, industry – I do not imply that these actors are united in their perspectives or that the offered quotes are 'typical' (see Crouch 2003a). Instead, I intend this research to offer glimpses into the material, sensory and cognitive practices of seed saving through a partial, situated, passionate retelling of the views and experiences that participants shared with me.

I was fortunate that several seed savers allowed me to visit them in their growing spaces. During these visits we walked around their garden or farm speaking about seed saving and related issues. Sometimes visits involved helping in the work by weeding, planting or cleaning seed, or watching children. Other times we just chatted. In many cases I met family and friends, and once a seed saver had arranged for us to be interviewed by the town paper to highlight the issue. Seed savers often demonstrated how they would save seeds, discussed why they had placed certain things together, showed me their stored collection of seeds, and told stories about their experiences and the seeds. Some gave me seeds they had been saving to add to my own garden and/or to pass along to others. In our discussions, savers' techniques and experiences blended with other issues including concerns about food, environment, family, neighbour relations, government policies, community priorities and corporate ownership.

These shared encounters with seeds and their savers in shared places offered insights into people's experiences and understandings of seed saving that might not have been apparent, or as deeply developed, otherwise. In their study of people–plant relations, Hitchings and Jones (2004: 9–10) note that '[t]he plants themselves were integral to the method as they prompted actions and conversations, and triggered a productive alignment of individual enjoyment and particular memory.' I think this observation extends to seeds as well. Handling, planting, sorting, inspecting, trading and eating seed during these visits raised questions, comments and stories. In this way, seeds made their own differences to this research. Further, these visits gave me an opportunity not only to see savers' techniques and practices, their collections and spaces, but also to share in their joys and frustrations with seeds.

How seed saving is enmeshed in other issues of concern is an important question. Offering a broader context to the perceptions and experiences of seed savers required collecting and analysing published documents. Included in this collection, oriented around international and national seed governance and contestation, were legal documents and governmental proceedings, industry reports and presentations, non-governmental research and proposals, media coverage and press releases. My analysis of these texts also informs this book, most clearly in Chapters 3, 4 and 5.

The results of an online survey of seed savers across Canada complements the primary research methods already mentioned. The survey was developed and promoted jointly between myself (hosted at the University of Wollongong) and two nongovernmental organisations – USC Canada and Seeds of Diversity Canada – that in 2012 initiated a collaboration on Canadian seed security. The survey was conducted in English and French, and received 342 responses (312 and 30 respectively). A sociodemographic summary is presented in Appendix 2. Participants responded to a total of 52 questions of various types including: dichotomous and multiple choice; multiple answer; Likert-scales; and open text, single or numbered spaces provided. Although the survey informs some of my thinking, I do not summarise all its findings here. Instead, select results and analysis add to other data in developing a fuller understanding of seed saving.

Reading This Book, or Chapter Outlines

Chapter 2 situates this study conceptually, working with the themes sketched above. This chapter helps connect the 'relational materialist' and 'multinatural' literatures upon which I draw with seed saving beginning with elaboration of practicing tomato seed saving, leading to a more theoretical discussion of lived experience and practicing as human-nonhuman relations. The second section considers the ethics and politics involved in attending realities as ongoing, practiced achievements, in which human and nonhuman agents participate. In this, I first take a closer look at rethinking nonhuman agencies and decentring humans, and proceed to ideas about worldly engagements – as intimate seed-saver relations and as encounters of friction among realities. This chapter is not intended as a synthesis or an attempt to 'build bridges' among diverse literatures or disciplines; rather, it offers some points of connection among formulations by particular theorists I have found helpful in understanding seed saving.

Following the idea of practiced orders existing in tension, Chapters 3, 4 and 5 explore seed-people orderings different from, but entangled with seed saving. These chapters consider how these orders are enacted and maintained, and address some ways in which saving, seeds and savers reconstitute and are reconstituted by these other worlds. In Chapter 3, I argue that seed-saver (and agrifood) realities are being reordered through processes of neoliberalisation in which transnational corporations are primary players. Emphasis in this chapter falls on the structure of this evolving order, as well as regulatory shifts, commercial market expansions, enclosure efforts, and revaluations. In considering the diverse neoliberalisation strategies of seed corporations, three cases are detailed: promoting certifying seed; dealing with contamination by genetically engineered organisms; and genetically engineering seeds. While the focus of this chapter falls on corporate, particularly agricultural biotechnology industry (ag-biotech), efforts to achieve dominant and durable arrangements, contestations and gaps in their efforts are noted.

Part of how seed corporations are able to reap high profits and expand control is through privatisation of seeds and exclusions of saving practices, using intellectual property rights (IPR). Chapter 4 is oriented around IPR. Seed-related IPR – plant breeders rights or patents – allow holders to restrict the use of a specified object, which in the case of seed might apply to varieties, organisms, parts of organisms (e.g., genes), or processes involved in breeding. I explore IPR policies, and proposed revisions, in the context of Canada in connection with international agreements and negotiations. As counterpoint to IPR formulations, the chapter evaluates farmers' rights as a concept and in its particular encoding through the International Treaty on Plant Genetic Resources for Food and Agriculture.

Chapter 5 explores the subject of genebanking. Genebanking predates the evolving corporate seed arrangements and environmental interests in biodiversity, and has its own relations with seed saving. Examining how it has evolved, how it now works, and its relations provides further depth to considerations of how seed-people worlds reinforce, exclude and/or reach-out. Though this might be seen as a slight detour, genebanking serves a dual purpose: conserving and making available plant genetic resources, particularly those useful for food and agriculture. Public and private breeders source material from genebanks, and there are reinforcing interactions – material and knowledge-based – among private capital, governments, international agreements and genebanks. As I investigate the challenges and options of genebanking though, it becomes clear that alternative possibilities exist, some of which are being sought out.

Chapters 6, 7 and 8 cover seed saving in more detail. Chapter 6 examines how and with who/what people learn to become seed savers. Following Bruno Latour's conceptualisation of 'learning to be affected' and Donna Haraway's 'response-ability', I explore the ways in which savers shape and are shaped by their practices with seeds (among other things). I pay particular attention to how savers cultivate and value sensory engagements with seeds and how they learn to attune themselves to living with seed, developing attachments to seeds and to saving in the process. Further, I elaborate on the experiential and experimental learning of saving, conducted as a collective collaboration among people and things, which results in ongoing development and embodiment of skills. The learning and attachments that take shape through saving open up possibilities not only for improved practice, but also for ethico-political engagements.

Chapters 7 and 8 further consider the ethico-political relations of seed saving. Chapter 7 centres particularly on seed-saver interactions, exploring saving as ethico-political praxis. This chapter develops two aspects of these relations: first, the valuations of savers of their practices as worldly engagements; and second, ideas of relational – always uneven – agencies among seeds and savers. Regarding worldly engagements, food provisioning and conserving biodiversity are investigated, while differential agencies brings discussion of seed-saver relations in selecting, adapting, accommodating and presenting. Val Plumwood's work in environmental ethics to decentre humans and allow nonhumans agency, and Donna Haraway's figuration of companion species, offer conceptual inspiration for this chapter.

The penultimate chapter furthers my argument that seed saving constitutes ethico-political practice. Building on the worldly engagements already considered in Chapter 7, Chapter 8 moves ethico-political engagement beyond seed-saver entanglements into how seed saving relates with other seed orderings, with particular attention to how savers see their engagements. In this chapter, seed saving is argued to constitute political engagement, not only in the sense of savers participating in official, institutionalised political processes but especially as they go about the everyday practices of saving seed. Using food politics to outline some of the possibilities and limits of analysing such everyday politics, I suggest that an approach that reads for diverse political engagements offers most insight into seed saving. Based on savers' comments, the chapter reveals ways in which savers, through saving, resist and contest other seed orderings while creating, fostering and cultivating alternative possibilities.

As a whole, the book carries a narrative thread, and might be read from end to end in the more traditional way. I hope a reading of this kind offers a steadily deeper understanding of seed saving encounters, experiences and meanings. Despite this, I do not think the chapters proceed necessarily in what might be considered a conventional structure – where each successive chapter rests on and further develops the ideas of those preceding. One might, for instance, if most interested in how seed saving is done, focus on Chapter 6, with additional sections from Chapter 7 and the first part of Chapter 2. Further, the chapters themselves play back and forth. For example, an understanding of seeds as plant genetic resources (for food and agriculture) is discussed as part of genebanking (Chapter 5), as part of the legal encoding of intellectual property rights (Chapter 4), and as part of savers' concerns in conservation and selection (Chapter 7), as well as serving as the basis for genetic engineering practices (Chapter 3). As a whole though, this book pays close attention to seed–people relations, with particular attention to seed saving, working to provide a fuller academic and practical understanding of the vitalities and importance of savers, seeds and seed saving.

Chapter 2
Rethinking Practice, Agency and Worldly Engagements

Fig. 2.1   **Practicing tomato seed saving**

The series of images above illustrates my tomato seed saving method. The steps included here might be given cursory descriptions:

1. Choose tomatoes from selection
2. Slice off tomato tops
3. Squeeze into container, add water (if necessary)
4. Label, wait for seeds to ferment
5. Rinse and drain
6. Spread to dry, wait
7. Test if dry
8. Label and store

But these images and steps present more than a how-to guide for saving tomato seeds – though they do that as well. This depiction of saving seeds serves as a way into a discussion of saving as an intimate encounter for savers (me) and (tomato) seeds, a practice to which we each contribute and one by which we are changed. Further, it provides opportunity to consider practice, practitioners and everyday engagement.

In this chapter I provide some elaboration of my approach to seed saving and several theorists and ideas that have recently informed it. In this regard, I argue for an understanding of practice as ongoing everyday relations among humans and nonhumans, which brings with it ethical and political implications and possibilities. The chapter offers, therefore, a section on practicing together and on ethico-political engagement, but first I consider what is missing from the above step-by-step method for saving tomato seeds.

What is Missing?

In the above depiction of saving tomato seeds much is lost. This lack is a reflection of partial – in the dual sense of incomplete and non-neutral – connections between me and the seeds, other savers, seed saving more broadly, and you as a reader (see Haraway 1988; Strathern 2004). Some of what I suggest is missing might be inferred, depending upon your own experiences, but either way my telling and the saving remain situated and, therefore, imperfect though productive. There are multiple ways in which I might explore this kind of partiality, but here I will highlight four points salient to ideas that inform seed saving as considered throughout this book.

First, in my depiction of saving tomato seeds some of the *experience* of practicing seed saving is missed. The rhythms – the moments of activity and waiting – for example, do not come through. Sometimes the seeds stick or I am tired/distracted/excited and saving is slowed/interrupted/bumpy, while other times we move together smoothly, quickly. There is also no sense that this seed saving started much earlier in the season as seeds sprouted, grew, adapted, reproduced and matured while I planted, tended, replanted, neglected, observed, marked, selected and resisted eating them. And what of the smell of the tomatoes fermenting; the feel of their pulps, juices and seeds squishing through my fingers; the challenge of running water into the container just to remove the fermenting liquid and unviable seed but not disturb the useful seeds; the satisfaction and skill of completing the task; the continuing vigilance to prevent bugs and mould; the fun and frustration of it all? In short, these images and instructions give an idea of some of what happens in seed saving, but much of the textures – the material, sensory, affective, emotive, cognitive aspects of everyday experiences are missed.

Second, the step-by-step portrayal ignores that both the seeds and I came to this encounter with our own *histories* – our own paths to 'here' – that influence our practice. My body, more attuned to academic tasks, takes a while to adapt to the process, but having done it for several seasons I have some familiarity from research and experience. I have been saving this variety of seeds for a few years – because their fruits are tasty, their plants grow well, and I am researching seed saving. They were given to me by one of the research participants from Québec after we chatted for some time at a seed fair. He had been saving them at his farm for several seasons, and was now selling them. As I save these seeds I am reminded of all the time invested this season and beyond it; that conversation and others; the gifting; my readings and 'how-to' workshopping. Having spent time with these seeds as they've grown for a few years, I know something about them and their lives, but there is plenty that I do not know – about past seasons with me and all the years before that. The experiences of these seeds are murky to me, but the seeds have embodied them and they show up in their forms and ways of living. Those involved in saving – including the seeds and me – did not appear out of nowhere. We have our own lived histories, developed capacities, and learned tendencies, and these alter and are altered by saving experiences such as the one outlined.

Third, primary attention here falls on the seeds and me – our intimate, if unequal, relations – rather than all the variously involved *others*. Some of these are mentioned in the steps – water, labels, jars, a knife. Then there is my journal and log book, other consulted texts, the weather, animals and insects, the storage closet, and so on – each with their own part to play, and each slightly different for each saver-seed collaboration. Coupled with these additional others are those omitted – evading light, warmth, moisture, bugs and disease is key to success. In mentioning these diverse presences and absences I mean not only to draw attention to the stuff of seed saving, but also to how it makes differences to seed saving as collective practice. The vitality and agencies of things deserve mention. I spend less time considering all the diverse things of saving (though see Chapter 6), but I return in various ways to the idea of seeds as material, meaningful and affective things.

Fourth, seed saving is a practice – a repeated activity – but it is *dynamic and differentiated*. Each enactment of seed saving reconstitutes a slightly different form, and yet it holds together, still recognisable as 'seed saving'. Saving with other tomato seeds in other seasons will not be exactly the same. Once you know the general approach, there are still details to figure out – and you must pay attention in different ways at different times in different places. Other savers have their own ways of saving with seeds – some don't slice the tomato first, ferment or label. Plus, saving lettuce, for example, involves different methods.

In this particular saving, not all the seed planted grew, not all the plants survived to bear healthy fruit, some seed remains in the garden and may grow next year without my interference, and not all the tomatoes were harvested for seed – ripe fruits were eaten by me, racoons, squirrels or other things, while unripe, overripe and damaged fruits became compost. Not all the seeds from each tomato fermented are saved – some became part of soup with their tomato pulp, were discarded as immature, were lost in the cleaning, and so on. I also check the seeds for marks, disease, 'poor form' and select carefully, eliminating some seeds while helping others continue.

Seeds continue because of and despite my efforts, and the seeds feed me, changing my tastes, nutrition and memories. Because the seeds and I are both dynamic and changing, so too are the ways we practice. Through our practices together we reconstitute ourselves and implicated socionatures. Rather than replication, saving is contiguous but progressive – seed saving in practice is a looping forward.

The images/instructions of saving tomato seeds and my notations of what is missed in their representation brings together several of my abiding conceptual interests – practices, food, activism, everyday life – with some more recent ones – embodiment, vitality, materiality, possibilities. The offered guide to saving tomato seeds and its missing elements, suggest ways in which these concepts connect with seed saving. In the next two sections, I explore two themes that collect some of these ideas and that orient much of my thinking. The first section deals with the idea of practice, considered particularly as an enactment among humans and nonhumans. This leads to the second section that details the ethico-political implications of such an approach for engaging in ongoing efforts to foster more responsible and sustainable relations.

Practicing Together

My approach to seed saving builds on an understanding of practice as patterned, but differentiated, interrelations among a set of bodies and understandings. As Reckwitz (2002: 250) puts it, a practice is 'a routinized way in which bodies are moved, objects are handled, subjects are treated, things are described and the world is understood.' Shove and Pantzar's (2005) examination of Nordic walking – speed-walking with specialised poles – is illustrative.

Shove and Pantzar explain Nordic walking, a relatively new practice, as a continuing and active integration of artefacts, images and skills; a process in which both consumers and producers are involved, and that varies with each country's supporting networks. They are careful to point out that it is not only external impositions that alter practices; rather, the components brought together vary and, as these things relate with people doing Nordic walking, Nordic walking itself is refashioned. Nordic walking continues and spreads only because people and poles persist in their collaborations:

If Nordic walking is to exist, people have to do it. More than that, what Nordic walking 'is' and what it becomes depends, in part, on who does it and on when, where and how it is done. It is in this sense that practices are inherently dynamic. (2005: 61)

A practice, then, is distinguishable, but differentiated, activity reconstituted by practitioners. In this sense, my tomato saving example above is just one instance of 'doing', which participates in a process of reconstituting seed saving as a recognisable, ongoing coming together of diverse elements. Shove and Pantzar's (2005) approach offers insight into assembling and continuations of practices, while other studies focus more particularly on the lived experiences of practices.

All of our practices are embodied and situated – whether it is seed saving, eating or something else. Much attention has recently been paid to experiences of activities such as walking (Middleton 2010; Wylie 2005), eating (Mann et al. 2011; Mol 2008; Roe 2006a, b), hunting (Lorimer 2006), cooking (Giard 1998; Heldke 1992), and even sitting (Bissell 2008). There is dissonance among the approaches – Wylie's nonrepresentational approach to a day's walk is slightly out of tune, for instance, with Giard's less personal sociological examination of cooking practices or Mann et al.'s (2011) anthropological ethnography of extending taste through bodies as people eat with their fingers. But each emphasises, among other things, the significance of everyday practices and lived experience.

Crang (2010) explains these attempts as moves to explore the textures – the senses, embodiments, affects and materialities – of everyday living. In a similar fashion, Carolan (2011) employs the concept of lived experience to 'enliven social theory by injecting living, breathing, feeling bodies into social methods and conceptual frameworks' to better think through the relations among material, sensory, and cognitive experiences of food. Both Crang and Carolan caution that studies of lived experience and everyday (seemingly insignificant) practices should not abandon concerns about private capital or socio-economics, a point to which I will return later. For now, I suggest that by highlighting the living-ness of particular practices this book shares with these studies a sense of the complexities of practices, illustrating how experiences affect participants and vice versa.

Practitioners not only shape their activities, but are shaped by them. Practice orients, disposes and reforms participants in material, sensorial, emotional and cognitive ways. In other words, through practice participants affect and learn to be affected (Latour 2004b; Lorimer 2007, 2008). People become Nordic walkers as they learn what the special poles allow them to do, as they train their bodies and minds to feel and understand this technique of walking, the landscapes walked, and themselves in altered ways. I explore seed saving as a collective learning practice in Chapter 6, but here I wish to note the more general point that in saving seed, seeds and savers reconstitute a practice that can be spoken about and drawn upon in subsequent enactments, while simultaneously being reconstituted themselves. If seed saving is understood as a continuing practice, then the ways in which savers speak about and engage with seeds are, at least in part, recreated through regular engagements. It is with this approach that I ask questions about how savers and seeds interact and are remade in practicing seed saving.

John Law and Annemarie Mol, together and separately, have for some years encouraged researchers to examine practices and it is from them I take a second example to explore how practices and their practitioners matter. Law and Mol (2008) begin not with people and poles, but with sheep in Cumbria during a foot-and-mouth disease outbreak. Law and Mol guide readers through veterinary, epidemiological, economic and farm-based practices dealing with sheep during an outbreak. These multiple ways of practicing biosecurity interrelate – reinforcing, conflicting or ignoring the other practices and the worlds they reconstitute. As Law and Mol explain, epidemiology's laboratory testing may deny a diagnosis resulting from veterinary practice; however, it may also be too slow in arriving, leaving vets to act on diagnosis without test results in efforts to pre-empt potential disease spread.

As practices are diverse and differentiating, so are the realities they recreate. As Mol (2002: 6) suggests, 'ontologies are brought into being, sustained, or allowed to wither away in common day-to-day, socio-material practices.' In Law and Mol (2008), each of the practices of biosecurity bring into being a particular ordering of sheep and people that remakes a world (or reality) – think of the business world's version of sheep biosecurity, or the reality of on-farm biosecurity work. Realities, or worlds, are enacted alongside practices, and multiple practices form multiple realities. To be sure, realities and practices are embedded in and sustained by broader social, political, natural and technological processes; but these processes themselves were and are made up of enactments. Accordingly, it is not only practices that are maintained through their repetition, but the realities that practicing assembles. Realities, bodies, thoughts, and so forth are all ongoing achievements that are made, remade, and even unmade through multiple enactments. Further, these realities work together; in the biosecurity example this means that biosecurity holds together or falls apart depending on how its multiple practices relate.

The same approach holds in considering those involved, in Law and Mol's case: sheep. Sheep are enacted in multiple ways. In veterinary practice, for instance, sheep become a potential host for the disease that requires diagnosis, but in farming sheep are (among other things) members of flocks threatened by the disease as well as by the protocols put in place to eradicate outbreak. This is not just an issue of perspective (i.e., I see sheep differently from you); rather, through differentiated practices sheep become otherwise. According to Law and Mol, the sheep is not one, but instead four (and more undescribed) sheep. These sheep versions, like the multiple practices described, reinforce and/or challenge each other, but all of them assemble as sheep-in-outbreak.

Connecting this insight about multiple enactments, practices and realities with seed saving might involve considering: multiple savings of different seed adding up to 'seed saving' practice; the dis/continuities of seed saving in comparison with other people-seed relations – genebanking or genetic engineering for instance; or the multiple practices and practitioners that come together in a corporate seed ordering. Each of these exemplars, and others, find themselves in this book, though some are more elaborated than others. A focus on practices, then, facilitates the task of exploring ways in which people and things affect and are affected as they relate, and how these practices – in their turn – shape and are shaped by the worlds accomplished through their enactments.

If practice is, in Schatzki's oft-quoted words, 'a nexus of doings and sayings' (1996: 89), then it follows that studies attend both what gets done and what is said about what gets done. To various degrees, studies of lived experience of everyday practices draw, as I do, on personal engagements and observation as well as other people's articulations about their experiences to try to get at the material, sensory, cognitive and affective aspects of practices. Storytelling, as vignette or narrative, evokes enacting worlds and can be effective in situating not only the practice but the researcher (see Lorimer, H 2010; Mann et al. 2011; Probyn 2011a; Wylie 2005). Further, some of this research combines this kind of approach with other practitioners' entanglements, as represented by explanations (see Carolan 2011; Giard 1998; Spinney 2006). This is important in highlighting how participants make sense of and value their practices, and vital to understanding why they do what they do. I use both these techniques, retelling my own experiences and observations as well as considering the words of practitioners with whom I spent time, to help me convey some of the experience of seed saving.

Ethical and Political Engagement

A wide range of studies now exists examining phenomena as dynamic, enacted ontologies involving a cast of human and nonhuman participants (for reviews see Anderson and Wylie 2009; Bakker and Bridge 2006; Coole and Frost 2010: ch.1). I find most suggestive for seed saving those analyses that recognise lively worlds, which we enact through engaging and learning with others – not all of which/whom are human. I have in mind here a diverse literature including vital materialism (Bennett 2001, 2010), ontological politics (Mol 1999, 2002), and more-than-human geographies (Whatmore 2002, 2006). You'll have read their influence in my language and analysis of the outlined gaps of the how-to example of tomato seed saving above. The differences among relational ontologies reflect their varied subjects, relations, and consequences as well as authors' dispositions toward relationality and ethico-political engagement. In contrast, these studies demonstrate shared commitments to understanding phenomena as founded in practices and processes, to decentring humans, and to dynamic, contestable worlds.

Studying practices, practitioners and their encounters is not only an exercise in celebrating complexity (though we may do this as well). There are also ethico-political questions and implications (Bingham and Hinchliffe 2008; Braun and Whatmore 2010; Coole and Frost 2010; Haraway 2008; Latour 2004a; Plumwood 2009; Stengers 1997; Whatmore 2002).

Attending reality as enacted and contested raises questions about how worlds and relations take shape, as well as about our choices regarding the worlds within which we participate. So, for example, we might agree that seeds are or contain genetic resources that can be privately owned through authorisations of intellectual property rights, and that corporate and government sectors are promoting or complicit in this particular approach to seeds. However, we need not accept that this approach is the only, or the preferred, means of understanding or relating with seeds. A number of ethico-political issues are raised by taking a view of worlds as practiced, two themes of which I highlight as salient to this research: taking seriously the liveliness and agencies of nonhumans; and, engaging in the world aiming to alter practices and realities in ways more responsible and sustainable.

Nonhuman Participation

First, relational ontologies shift humans from their position as privileged actors. Braun and Whatmore (2010: xix) insist that, 'nonhuman and technical objects are an irreducible part of the stories of the becoming-being of the human, both individually and collectively, that this could not be otherwise.' In this insistence, they challenge analysts to consider more-than-human politics and agency, arguing that humans, like other bodies, are already caught up in a lively world. Taking nonhumans seriously, therefore, requires attention to their roles and enrolments.

At one point in her philosophical work considering nonhuman involvements Bennett (2010: 38) illustrates the limits of human determination within a particular assembling with a short reference to riding a bike on a gravel road. She explains that '[o]ne can throw one's weight this way or that, inflect the bike in one direction or toward one trajectory of motion. But the rider is but one actant operative in the moving whole.' As bike, road and rider are in bike-riding together, Bennett's example also shows how particular nonhumans – roads and bikes – work with and against a rider to accomplish riding. The rider must not only adjust to nonhuman influences, but will develop corporeal sensibilities in bike riding – how the feet are placed, the back bent, the handles held, and so on, is all learned through experience and embodied. In undertaking a practice therefore, it is not just that a person uses things to attain particular ends; rather, people and things work and change together.

In rethinking nonhuman agencies, Bennett uses the phrase 'thing-power' to refer to 'the curious ability of inanimate things to animate, to act, to produce effects dramatic and subtle' (2010: 6). This has something in common with Albert Borgmann's (1995: 90) contemplation of what he calls focal things, in which he argues that such a thing 'has an intelligible and accessible character and calls forth skilled and active human engagement', and further that 'the experience of the thing is always and also a bodily and social engagement with the thing's world' (1984: 41). While Borgmann's formulation might be considered somewhat essentialist in that focal things possess characteristics, recent thinking on things shifts discourse to more relational agencies. In *Vibrant Matter*, Bennett makes such a move, clarifying that 'thing-power' is perhaps too strong a term for what she intends, overstating the stability and independence of things. She instead rethinks things through capacities, vitalities and assemblages, which she feels convey a more dynamic and relational understanding of things and their interactions. The vitality to which Bennett refers includes 'the capacity of things – edibles, commodities, storms, metals – not only to impede or block the will and designs of humans but also to act as quasi agents or forces with trajectories, propensities, or tendencies of their own' (2010: viii). Whether considered as expressing power, providing focus, or acting on tendencies, 'things' here gain some agency – something often reserved for humans.

Among advocates of decentring humans there is tension around the fixity and individualism of things, and about how agency is expressed – as Bennett's shift in language suggests. Jones and Cloke in their study of trees caution against approaches going too far with relationality, risking becoming 'insensitive to complexity and heterogeneity' (2002: 49). In their view, too much emphasis on relations may obscure the differential agencies and forms of individual organisms – such as a tree. A similar risk can be read within theories of practice in which participants become 'carriers' of a practice, rather than active shapers of such practice (see Reckwitz 2002).

Acknowledging this risk but working to retain the relational and practiced approach, concepts have altered from describing essential characteristics possessed by individual, human agents who act on things, to recognising capacities, affordances, dispositions, tendencies and/or orientations that emerge only through practice among differentiated others (see Ahmed 2010; Coole 2005; Ingold 2000, 2010; Latour 2005; Lorimer 2007). As examples, in this sense, corporations, rather than possessing power and control, have the capacity, under the right conditions, to exploit the opportunities offered by intellectual property rights to privatise seeds; savers accomplish seed saving only because seeds afford the possibility through their capacities to reproduce and adapt; eaters are disposed to particular foods because of previous experiences with them.

Recollecting Law and Mol's (2008) discussion of sheep in biosecuring, the sheep are not only enrolled by people's practices in particular ways; sheep are not totally passive in biosecurity. Instead, sheep engage, in sometimes predictable and disciplined ways but also in surprising and stubborn ways. As they interact with people enacting biosecurity, sheep may acquiesce or panic, roll and bleat, resist diagnosis, spread disease, defy models, display symptoms, suffer confusion, attract controversy, and so on. This does not mean that they control processes, or that they have the same agencies as those around them, but as sheep interrelate they change the possibilities of biosecuring foot-and-mouth disease. Sheep, in this sense, act.

Decentring humans evokes questions about who/what enacts change, and how. Law and Mol's (2008) sheep, Shove and Pantzar's (2005) poles, Bennett's (2010) bike and road are but three examples of a diverse and growing literature exploring phenomena through decentred and relational approaches. Agency in these works is redefined as an ability to bring about change or action, and humans are not the only ones that make a difference. I want to be careful here to clarify that while there are many entities involved in reconstituting realities, not all participate in the same ways or change relations to the same degrees. Relations are permeated with power.

Acting is not a choice of complete control or nothing, instead a wide range of agentic engagements is possible. Latour (2005: 72) explains '[in] addition to "determining" and serving as a "backdrop for human action," things might authorize, allow, afford, encourage, permit, suggest, influence, block, render possible, forbid, and so on.' Haraway (2008: ch.3) uses 'degrees of freedom' to capture the relative capacities of different bodies to influence and be influenced, making it clear that no one/thing is ever unencumbered or totally controlled – humans and nonhumans alike live together with relative degrees of freedom. There is some parallel here with arguments that domination or worldly order is never complete; instead, acting for change is always possible, though it is constrained to various degrees. In this sense, both humans and nonhumans engage in diverse, always unequal, ways.

In multinatural approaches – relational ontologies in which diverse living beings provide focus – much attention has been paid to animal-human associations, particularly to interrogating similarities and differences, boundless and situated bodies, and intimate – though not always kind – relations. It is, I think, easier to see nonhuman agencies when looking at sensate, living creatures – like sheep. But efforts to decentre humans and distribute agency have been variously explored in relation to all sorts of things – from technological artefacts (Bijker and Law 1997; Braun and Whatmore 2010; Latour 1991;) to plants and animals (Haraway 2008; Jones and Cloke 2002; Whatmore 2002; Wolch and Emel 1998). This growing body of work accepts nonhumans as potential agents with their own lives, capacities and tendencies.

Though less common, plant-human studies are appearing, questioning assumptions that come with limiting attention to animal's agencies (Hall 2010; Head and Atchison 2009, 2012; Hitchings 2003, 2006, 2007; Jones and Cloke 2002; Marder 2012; Power 2005; Schneekloth 2001, 2002; Staddon 2009). Plants seem even further than animals from human experience, more dissimilar to our sense-making in comparison to animals. And yet people–plant relations are just as – perhaps more fundamentally – mutually shaping. Plants and people breathe together, we use each other as nutrients, we kill them as they escape our inevitably failing controls, we conserve and cultivate them as they ambivalently respond with flourish or indifference.

Seeds have also garnered some attention – particularly in disguise as food, genetic resources, and genetically engineered technologies, but also as they travel and are grown (Clark 2002; Fowler and Mooney 1990; Nazarea 2005; Shiva 2000; van Dooren 2007, 2008, 2009; Whatmore 2002: ch.5, 6). Whatmore (2002: ch.5, 6), for instance, explores plant genetic resources and genetically engineered foods. She offers a compelling analysis of seed-people interrelations within agrifood processes, and in each form seeds shape and are shaped by their people and technological and political relations. But Whatmore's analytic interest, like that in most seed-related studies, does not dwell on practices of seed saving. This book attends seed saving, attempting, as part of that project, to take seriously experiences and agencies of savers and seeds.

In line with these approaches, I choose to understand people and things as more emergent than essential, but with acknowledgement that all bodies – human and non – are accomplishments with tendencies and capacities already shaped through previous experiences. Seeds and savers come with their own histories and present particular opportunities for enactment together. As Haraway (1997: 89) indicates, '[s]eeds are brought into being by, and carry along with themselves wherever they go, specific ways of life as well as particular sorts of dispossession and death'. This means rethinking savers' centrality as well as seeds' potential. People-seed relations and agency manifest in seed saving are more fully considered in Chapter 7, particularly in relation to Plumwood's challenge to decentre humans and Haraway's idea of companion species. This theme runs throughout the book as the capacities of seeds to ignore, elude, embody, facilitate, and so on, provide support and contest to worldly arrangements.

Further, understanding seed agencies is only part of the contribution decentring humans offers in considering seed-people relations. Intellectual property rights, for instance, have been conceived as legal devices performing and substantiating proprietorial assemblies, making practical and material differences in the bodies of seeds, the senses of savers, and their possibilities of acting together (see Carolan 2010, Whatmore 2002: ch.5). So although much of this book's attention falls on seed-saver relations and encounters, other things –

intellectual property rights, genebanks, bacterial DNA, policy proposals – make appearances, shaping and being shaped by seed saving.

Worldly Relations

There are differing levels of commitment and varied stances to considering ethico-political implications, but increasingly normative calls echo in relational ontologies as researchers explore ways of 'living with' ontological difference – looking for more responsible and sustainable ways of engaging in and reconstituting shared worlds. Much of this work, building upon insights of phenomenology, environmental ethics and feminist analyses, begins with 'an embodied and practically engaged self ... from what human beings do in the world ... so as to rediscover the totality of practical bonds with others' (Kruks 1995: 11–12 as in Whatmore 2002: 154). In this way, ethics becomes something other than abstract, universal imperatives of right or wrong, and politics extends beyond official institutions. Through situated engagements, ethics and politics become about 'responsibility and accountability for the lively relationalities of becoming of which we are a part' (Barad 2007: 393). Ethics and politics, in other words, become about everyday practices.

The first move in such a project involves engaging with others and their worlds, learning to recognise and appreciate difference. As Bennett (2010: 14) suggests, part of '[t]he ethical task at hand here is to cultivate the ability to discern nonhuman vitality, to become perceptually open to it.' In exploring this idea in relation to everyday practices, several social studies scholars have taken up Latour's (2004b) concept of 'learning to be affected' (Gibson-Graham and Roelvink 2009; Hinchliffe et al. 2005; Lorimer 2008). I explore seed saving in relation to this notion in Chapter 6, adding to this literature. As Latour (2004b) explains, to discern difference involves training – or practicing. Embodied and situated skill development, whether through formalised training or informal practicing, provides opportunities to affect and be affected by others and their worlds. In this way, 'in learning to be affected – in articulating propositions – bodies, things, and words all have the potential to become more than they were before' (Hinchliffe 2003: 216). Further, as practices and others are complex and changing, so must be our efforts to learn them. Attuning, therefore, requires and results in layered engagements – physical, sensorial, cognitive. It takes practice to cultivate capacities to sense and make sense of others and their worlds.

Though it is imperative that we respect difference, and learn through our engagements how to do this better, we can never truly know the other or settle our recognitions of their differences (Haraway 2003; Latour 2004a). This idea of never finalising knowings – or always partial connections – is fundamental to practice-relational approaches. It opens opportunities for change, but also demands our continued engagements and reconsiderations.

It has been recognised for some time that knowledges are situated and enacted, a point developed by Haraway (1988) to critique knowledge purporting to be universal and neutral. But in her development of situated knowledge Haraway went further, arguing not only that knowledge is partial, located and ever-changing, but that some knowledges are better than others, and that all carry with them ethico-political traditions and implications. Following Haraway, part of my interest lies in examining the practices of seed saving, the knowing that develops through these practices, and how this practiced know-how may offer 'something quite extraordinary, that is, knowledge potent for constructing worlds less organized by axes of domination' (Haraway 1991: 192).

If part of the ethico-political task is to discern differences and vitalities (human and nonhuman), the other part has to do with using this learning to foster better ways of coexisting. Haraway's (2008) recent work on companion species, for instance, develops an ethics of shared suffering and response-abilities in which curiosity, learning, engagement and empathy inform practice with nonhuman others. While Haraway relies upon a shared sensitivity more reflective of human–animal possibilities than those of people–seeds when discussing suffering, her insights regarding responsibility and degrees of freedom are useful in rethinking seed saving. In Chapter 7, I explore savers' relations with seeds as ethico-political relations, in part moved by Haraway's approach. Haraway focuses on species, following a rich tradition of animal studies, but others have considered a range of things including everything from the earth (Clark 2011) to micro-organisms (Hird 2009), not to mention a plethora of science and technology studies dealing with nonliving artefacts.

In a different philosophical vein, Bennett (2001, 2010) attempts to locate ways of recruiting people into finding better ways of living with others. She finds hope specifically in wonder-filled attachments. While wonder carries with it a threat of spectatorship, Bennett insists on material, practical engagements offering as examples encounters with such things as potato chips, electrical outages and landfills. These examples suggest Bennett's interest is not in a wonder that is only joyful, but rather one that combines disturbance with delight, that involves surprise and hesitation. Elaborating on wonder, or enchantment, as a motivating affect, she argues that it 'points in two directions: the first toward the humans who *feel* enchanted and whose agentic capacities may be thereby strengthened, and the second toward the agency of things that *produce* (helpful, harmful) effects in humans and other bodies' (2010: xii). Again, we see why rethinking agency is crucial to such an approach. Further, the role of emotion and relation is here highlighted. How to recruit and retain savers as practitioners is an important question if we hope to keep saving practices, and the seeds that come with them, alive. The roles of affect and attachment matter in such an effort.

Ethico-political engagement requires ways of rethinking politics that account not only for nonhuman difference and vitalities, but approaches to politics. For Mol (2002) rethinking politics to involve questions of practice means examining existing practices, finding ways of contesting seemingly unavoidable or unitary realities, pointing to ways in which realities do (not) cohere, and highlighting how worlds might be (re)made otherwise. In other work considering potential change, Gibson-Graham rethink economic arrangements and analyses. They endorse a strategy of 'starting where we are' and piecing together understandings of living differently as part of more-than-human communities by asking 'what is being done' to create alternative economies of non-capitalist development (Gibson-Graham and Roelvink 2009). There are, they suggest, 'glimmers of the future' in ongoing projects, which in all their diversity might be enrolled in constructing an economy 'more focused on social wellbeing and less on growth and profitability' (Gibson-Graham 2011: 2). What, they ask, can we learn from practices already in trial?

I follow Mol's and Gibson-Graham's challenges to learn from what is being done already in my consideration of seed saving and the possibilities it presents. Saving is, in part, a response to other worldly relations of seeds and people. As an everyday ethico-political practice, savers with seeds respond to other seed-people realities – an idea explored throughout the book, but especially in Chapter 8. Whether theorists emphasise response-ability, wonder, multiplicity, diversity or something else, more are posing questions about what it would mean to our ethico-political relations to take nonhumans more seriously and how to foster better realities-in-practice. This book contributes to that effort.

Recognising a lively material world in which we learn through our engagements brings with it ethico-politics for the practice and practitioners being studied, as well as for the researcher (Greenhough 2010; Haraway 1997, 2008; Whatmore 2002). All practices – research included – change realities. Law and Urry (2004: 404) explain that, in acknowledging practice in this way, the question becomes which realities 'do we want to help to make more real, and which less real? How do we want to interfere'? For researchers this kind of approach means becoming passionate, modest witnesses, caught up and intervening in our studied worlds – whether we like it or not.

Considering Conflicts

Two related cautions for the range of ethico-political engagements discussed in this section are worthy of mention here – each of which highlights discontinuities and conflicts. First, though nothing in 'living with' accounts suggest that conflict is absent or that cohabitation is easily achieved (see Bingham 2006; Hinchliffe 2010), there has been a tendency to highlight more comfortable and cosy relations than ones that confront, exclude or are absent.

Not all relations are friendly; we eat, use, ignore and kill nonhumans – including seeds – in addition to care-fully attending them in our practices together. Considerations of animal welfare in food production and laboratory environments (Greenhough and Roe 2010, 2011; Haraway 2008: ch.3, 10; Miele 2011), climate change (Yusoff 2009, 2010), and biosecurity (Beisel 2010; Greenhough 2012; Phillips forthcoming) offer insights into difficult, ambivalent and/or deadly relations. In addition, as Clark (2011) reminds, reciprocity among those involved is not assured – indifference to human presence and influence is part of our always already lively worlds. Acknowledging nonhumans who/which have lives of their own means allowing that humans are sometimes ignored and excluded in their own turn. This does not excuse people from engaging, rather, the question becomes one of when and where human-spurred interventions may be most strategic.

Seed saving makes significant, even strategic, ethical and political differences. Though seed saving is, on the whole, a practice in which seeds and savers come together, it is not without its difficulties, exclusions, indifferences and losses. By attending some of these in addition to more pleasant aspects we gain a fuller understanding not only of lived experience and particular phenomena, but of motivations for engagement. Just as wonder (Bennett 2001, 2010; Ellis 2011) and *jouissance* (Lorimer 2008) have their parts to play, loss and mourning can be motivational and productive (Yusoff 2010). Moreover, 'accountability and responsibility must be thought in terms of what matters *and what has been excluded* from mattering' (Barad 2007: 394, emphasis added). Inclusions bring with them exclusions – I save this seed and not others, this seed gets marketed and others do not, and so on – and these vital relations matter in ethico-political practice.

The second related caution extends the first, moving it beyond consideration of bodies to worlds. Worlds and beings emerge, say Anderson and Harrison (2010: 8), through 'active, productive, and continual weaving of the multiplicity of bits and pieces.' Here, 'world' refers not to the earth, but to an ensemble of practices that hang together forming a context in and through which particular relations, senses, things, and so on come into focus. I use the term worlds – along with realities and orders – in a similar way, asking about seed saving worlds in which seeds, people and all their commensurate others learn, collaborate and respond. Recalling Law and Mol's study of biosecuring sheep, a world holds as multiple practices, themselves assemblies of diverse elements, come together in its reconstitution (see also Law 2002; Mol 2002). Further, just as there are multiple enactments of any reality, there are multiple realities.

Understanding worlds as ongoing enactments opens up possibilities of practicing realities differently – of ordering things in other ways. But the promise of always enacted worlds does not suggest that worlds are totally open, or that all paths are possible – orders may dominate, worlds endure, realities fade.

Enacting different relations is possible, but not necessarily easy. If there are multiple worlds, what happens as these worlds encounter each other? In supportive relations, a more durable order evolves, but what about in collision or conflict? And what of when realities just ignore, without confronting or reinforcing? As Hinchliffe (2007: 187) notes in an examination of the interactions of a garden multiply practiced, '[a]ttempts to enact a singular version of economy or nature can have the effect of making some modes of ordering, like care, less visible.'

Ordering a world in one way, in contrast with alternative possibilities, constitutes ethico-political engagement – though not always for the better. Again, attending to difficulties, discontinuities and exclusions is necessary. Moreover, relations may be more complex than the options commonly presented: support or oppose, perhaps ignore. How savers respond to other seed orderings through their practices is considered particularly in Chapter 8, though connections of other worlds with saving, savers and seed constitute part of each chapter.

How to address orderings and powers without ignoring dynamism and lived experience is a continuing challenge for researchers. In her own effort, Tsing (2005) offers what she calls an 'ethnography of global connection.' Drawing upon fieldwork within communities of South Kalimantan, Tsing (2005: ix) stresses that she tells:

> a story that is insistently about social nature in a particular place. Yet simultaneously: A story of North American investment practices and the stock market, Brazilian rubber tappers' forest advocacy and United Nations environmental funding, international mountaineering and adventure sports, and democratic politics and the overthrow of the Suharto regime, among other things.

To do this, she moves among multiple places with multiple species, concentrating on relations of friction. This idea of friction – of rubbing up against – is useful in understanding how seed saving practices in yards, farms, community gardens, and even road medians are intimate stories with global connections, existing alongside and sometimes wrapped up with processes of conserving, breeding, corporatising, privatising and so on. It is also useful for considering how, for instance, more institutionalised and official political engagements meet everyday ethico-political praxis. Other work including Whatmore's (2002) studies of hybrid geographies, Haraway's (2008) rendezvous with companion species, and Probyn's recent journeys with tuna (2011a) and kangaroo (2011b) illustrate how politics, economics, ecologies and ethics are not separate realms or abstractions, but are partially connected with material, sensory, cognitive experiences of humans living with nonhumans.

In following what is being done to and with seed saving, I am led to consider practices and orders that contain and constrain the practice and refigure its practitioners. Seed saving encounters multiple seed-related realities, a few of which are explored in following chapters – conservation in genebanks, privatisation and concentration in agri-business, interference through gene technology and governance through intellectual property rights and standards. In these critical engagements, I add to the literatures already explored in this chapter, as well as others that favour examination of neoliberal natures and agrifood networks and that remain open to relational ontologies and lived experiences. Reviews of neoliberal natures (Bakker 2009, 2010) and agrifood scholarship (Cook et al. 2013) indicate that each field would benefit from relational, entangled, affective accounts of lived experiences, and I think the reverse also applies.

In light of various ecological and agrifood crises, considerable attention is being paid to accelerating commodification of natures, increasing involvement of private corporations, and shifts in governance as part of neoliberalising processes. This literature on neoliberalising natures builds on strong traditions of political economy and political ecology to examine the management of resources within a neoliberal capitalism. In her review of resource studies, Bakker (2010) continues previous work (see Bakker and Bridge 2006; Bakker 2009) by endorsing an approach in which natures are understood to be more than only resources, agency is distributed, and neoliberalisation is variegated. Acknowledging exceptions, Bakker juxtaposes her suggestion of a more relational ontopolitical approach with trends, especially in political economy, to consider nature as a source of potential commodities and, more generally, to adopt human-centred positions.

Recognising that neoliberalisation manifests differently, much of this literature of neoliberal natures pursues case studies of particular natural resources as commodities and their associated neoliberalisations. However, in recent reviews, Bakker (2010) and Castree (2010a, b, 2011) each outline a range of strategies that might manifest in particular instances of neoliberalisation. Each list differs, but there are overlaps regarding privatisation, marketisation, deregulation, reregulation, market proxies, and the cultural and political alienation of people. Following Bakker and Castree, neoliberalisation is a contested process that unfolds as a range of strategies (such as privatisation), that vary with the target (like seeds) and the particular form (such as patents). Later chapters more specifically consider private capital's efforts to expand markets and deepen commodification through strategic technopolitical arrangements. Particular cases in Chapters 3 and 4 illustrate changing regulations, attempted subsumption through genetic engineering, and privatisation through intellectual property rights. The refigurations of seeds, savers and seed saving that come with these processes are also explored.

If neoliberalising natures can be said to orient around particular natural resources as commodities, a similar approach is employed in food studies. An identified gap between production and consumption has led to popularity of 'following the thing' or investigating commodity chains (see Belasco and Horowitz 2009; Goodman and DuPuis 2002; Lockie 2002; Whatmore and Thorne 1997). Following the thing – food or otherwise – involves investigating the specified thing's journey, and its transformations and valuations along the way (Cook et al. 2006). It serves also to relate food journeys to 'the norms, practices, and social institutions shaping, and being shaped by, relations between actors in the commodity network' (Freidberg 2004: 10). So popular now that it might be considered its own genre, Whatmore and Thorne (1997) employed this kind of following technique early on in an exploration of fair trade, demonstrating globalisation as a contested process and alternative food networks as offering diverse resistances and possibilities. Whether following foods or other commodities, insights are gained by discerning and critically engaging with things, their orderings and respective impacts.

In a recent review of food geographies, Cook et al. (2013) suggest adding to 'follow the thing' studies in two ways: first, with considerations of more entangled relations than are usually portrayed with the typically linear approach of food chains; and second, with engagements conveying more of the material, affective, 'living with' aspects of food relations. In this book I do not follow a particular seed – a GE canola, heritage tomato, or triticale seed for instance – as I might in a more straightforward application of this approach. However, this research has some kinship with these studies in that I attempt to follow seed saving – and its practitioners – into diverse spaces and refigurations, and in considering seed saving as lived, entangled, worldly relations.

Summing Up

Getting back, then, to everyday practice. Some considerations of everyday practices maintain humans as privileged – if constrained – actors (Bourdieu 1977; de Certeau 1984; Foucault 1990; Lefebvre 1984; Schatzki 1996); although nonhumans – living and non – may influence outcomes, they remain objects used (or not) by humans. However, the more inspiring and provocative approaches discussed in this chapter decentre humans, arguing for more distributed agency. Further, where some studies might be accused of essentialism, the work outlined here is (to greater and lesser degrees) relational. No body holds characteristics or power as isolated, unitary selves; rather, each has their own changing capacities, affordances and dispositions that emerge as multiple bodies encounter one another. Since it is through the practices of people and things that worlds emerge, these worlds are continually changing (some less than others) and therefore, change is always possible. The future is uncertain, undetermined, and may be made better (or worse). Everyday practices, therefore, are immersed within and responsible for dominant orderings, but they also present opportunity for changing relations to reconstitute worlds in better ways.

My enactment of tomato saving, mundane and momentary as it was, participates in wider worlds. It is part of ongoing seed saving practices that reconstitute seed, savers and their realities. It is only because of previous savings that those tomato seeds allowed me to save them for another season's planting, and in re/enacting saving I work to maintain that possibility for others and all it brings with it.

If how we engage in our always already-lively worlds makes material and ethico-political differences, as this chapter has suggested, then examining how and which options are maintained, opened and shut down through various seed-people relations is valuable. Attention to seed saving and its practitioners remains limited in social studies, and this book goes some way to addressing that lack. How seed saving is accomplished, and what else is being achieved through the practice is explored in Chapters 6–8, while the next three chapters consider seed-people relations other than saving, asking how those realities take shape and how they relate with savers, seeds and seed saving. These issues are perhaps especially important now, when engaging in seed saving is becoming increasingly difficult as other seed orderings take over, but when the practice persists in offering opportunities to rethink and reform future ways of living with seed.

Chapter 3
Reordering with Corporations

'Monsanto Completes Acquisition of Seminis' reads the headline. The press release from Monsanto's media office is five lines long, including the 'for more information ...' line (Monsanto 2005a). The brevity of the press release belies the event's significance. With this purchase, Monsanto – already a major biotechnology and agricultural chemical producer – became the world's largest seed company, gaining control of significant market share in fruit and vegetable seeds. At the time, Seminis supplied around 3,500 varieties to businesses world-wide – especially North America and Europe – and an estimated 40 per cent of US vegetable seeds (ETC Group 2005; Monsanto 2005a). Seminis varieties were and are used by all kinds of farmers – organic and chemical, small and large scale, low and high tech. Seminis exists in many growers' fields and on many consumers' plates.

Despite this presence, Seminis is not well known to growers or eaters. At the time of the purchase, Seminis seeds were found alongside those from other suppliers in multiple seed catalogues – Burpee, FedCo, Johnny's Select Seeds, Stokes, among others. But catalogue seeds are known by variety rather than original supplier, so few recognise the seeds as Seminis's and may believe that the company they buy from grows their catalogued seeds. Monsanto's purchase received little attention in popular media, but it made waves in seed circles. It presented some tough choices for those affected. Gardeners, farmers and retailers took positions along a continuum – at one end were those concerned enough with the politics and potential impacts of the sale to drop varieties immediately, while at the other were those more interested in variety availability who chose to continue with Monsanto-Seminis (Ball 2012; Dillon 2005; cf. Fedco 2005). Even for those concerned about the sale, for some crops and in some places options for quality seeds and varieties were limited. Monsanto's purchase of Seminis was finalised in March 2005, but it is suggestive of ongoing concerns about corporate control and concentration, and the implications of those processes for seeds and people.

This chapter develops the argument that seeds and savers increasingly encounter a technopolitical reordering of seeds – one reconstituted through processes of neoliberalisation in which corporations are the primary, but not sole, actor. Collaborations between governments and seed corporations facilitate commercial market expansion and regulatory constraint of seed saving, while new breeding techniques allow unprecedented exploitation of the capacities of seeds for creating new commodities. What I refer to as the corporatising (or corporate) seed order is a complex achievement of many actors employing multiple strategies to adapt, expand and make more durable this reordering.

The influence and importance of the corporate seed order needs to be attended, while acknowledging that its consolidation remains incomplete and contested.

Following a broader consideration of this corporate seed reordering, I present three examples of strategies to expand markets, revise regulations and deepen commodification. The first combines altering and repurposing regulations to increase profitability through revaluing saved seed while promoting certified seed. The second investigates an event of contamination of Canadian-grown flax by genetically engineered organisms. I also consider the implications of this contamination and subsequent arguments about assuring markets – through acceptance of low-level presence or economic impact evaluation. The final example explores the use of biotechnology as a means of furthering seed commodification, with particular attention to Terminator (or genetic use restriction technology) seeds. Each case illustrates manifestations of the corporate seed order as evolving through neoliberalisation processes, as well as contestations of this reordering. In the final section, I reflect on implications of the corporate seed order for seed and saver refigurations as well as some remaining fissures.

A Corporate Seed Order

Increasingly, the food we eat traverses complex and international agrifood chains, at least part of which are controlled by only a few transnationals (FAO 2003). Food and agricultural products have long been part of international trading and power plays, but over the last 50 years agrifood production, distribution and marketing have become increasingly global (Magdoff et al. 2000; Weis 2007). In Canada, a different set of three or four transnationals dominate each sector – agricultural inputs, outputs and retail. This situation prompts one farmers' union to describe the situation in this way:

> A customer puts $1.35 on a grocery store counter for a loaf of bread. Powerful food retailers, processors, and grain companies take $1.30, leaving the farmer just a nickel. Powerful energy, fertilizer, chemical, and machinery companies take 6 cents out of the farmer's pocket. Taxpayers make up the penny. (NFU 2005: 10)

In addition to control within agrifood chains, corporations assume influence in governing, involving themselves in lobbying, private standard-setting, implementation and policy-making processes (Clapp and Fuchs 2009). The economic and political strength of farming communities once gave them influence in Canadian politics, but as economics have shifted government is less able, or less willing, to support farmers against industry interests (Clark 2005; Mooney 2006a; Wells 2005; Winson 1993).

The reordering of agrifood in the last 50 years provokes questions about its character, maintenance, effects and contestation. McMichael (2005, 2008, 2009) and Friedmann (1993, 2005) each analyse periods of relative stability and transition in world agrifood arrangements, arguing that a new era is emergent – one characterised by neoliberalising processes in which corporations are central. McMichael's analysis, oriented around global South dispossessions, illustrates ongoing conflicts between an agriculture that is increasingly export-oriented and corporate-controlled, and countermobilisations of food sovereignty movements. Situating her analysis within the global North, Friedmann suggests that there are emerging tensions between whether our agrifood networks will 'be mainly private and corporate, or public and democratic' (1993: 52). With waning American influence and shuffling governmental priorities, Friedmann (2005) argues that, as it consolidates, this emerging order shifts historical arrangements through privatising regulations and control, exacerbating existing class differences and deepening commodification. McMichael and Friedmann share conviction that our agrifood orders are neoliberalising with corporate entanglements, and that resistance persists in mobilisations for socio-economic and ecological justice. I take inspiration from their analyses to consider the ordering of seed, proposing that seed is undergoing similar rearranging through neoliberalisation led by corporations, and that this corporate seed order is not uncontested or unchangeable despite appearances.

The contemporary seed order emerges from shifts beginning in the 1970s, when several technopolitical changes came together. In the mid-1970s in a context of wider agrifood changes, small biotechnology companies took advantage of state subsidies to pursue development of newly discovered genetic engineering techniques and their numbers, budgets and options grew (Andrée 2007). Also at this time, corporations formed alliances with government to alter public breeding practices – dominant since the late 1800s – gaining new access to laboratories, research, and materials, while reinforcing marketisation (Kuyek 2004b; Moore 2002). In addition, agri-chemical corporations – faced with chemicals coming off-patent and high costs of new product development – became excited about the potential of genetic engineering for commercial seed, and bought out thousands of smaller seed houses.

A frenzy of mergers, acquisitions, spin-offs and collaborations among seed, agri-chemical and pharmaceutical corporations occurred in the 1980-90s, improving corporate positioning in markets and governance. In the mid-1990s, the first genetically engineered (GE) seeds hit the North American market and, with supportive governments, agricultural biotechnology corporations (ag-biotech) thrived. Combining biotechnologies with legal protections increased profitability and privatisation in the seed sector, driving market consolidation. Vertical and horizontal integration strategies consolidated production processes as well as crop and input sectors, restructuring the seed industry further (see Figure 3.1). Today, neoliberalisation continues its advance with corporations pursuing ever stronger technological and political control over seeds and savers, facilitating market expansion and profit increases.

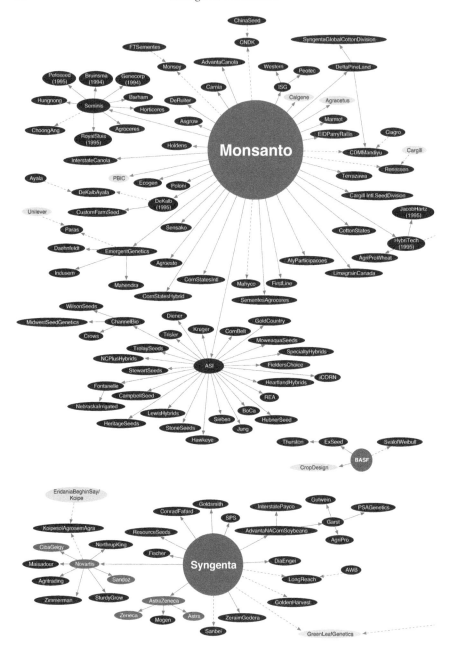

Fig. 3.1   Seed industry structure 1996–2008

Adapted from: Howard 2009

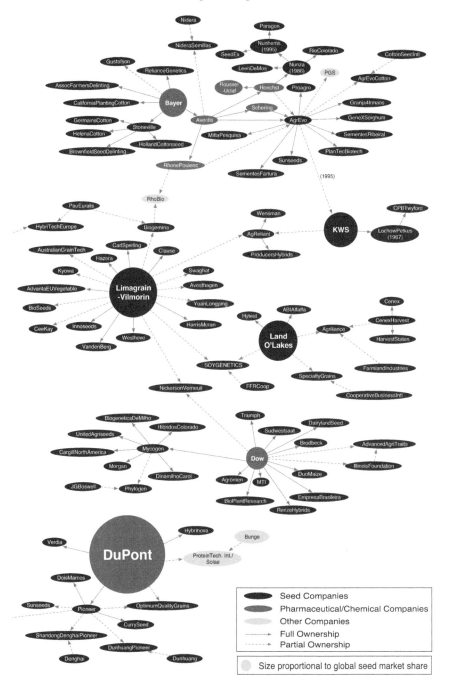

Seed Companies
Pharmaceutical/Chemical Companies
Other Companies
— Full Ownership
----→ Partial Ownership

Size proportional to global seed market share

Consolidating Control

Commercial seed markets are growing in value at both national and global scales. Global commercial seed amounted to USD21 billion in 2004, compared with 27.4 billion in 2009 (ETC Group 2005, 2011),[1] and Canadian industry estimates seed sales at CAD2.2 billion in 2009, compared with 1.5 billion in 2004 (CSTA 2004a, 2011a). Canada's seed imports recently overtook exports, with a negative balance of trade in 2008 reaching an estimated CAD95.7 million in 2009 (CSTA 2011a). So, the value of commercial seed markets is growing, and while the domestic seed industry does not always benefit, transnationals are prospering.

Commercial markets are also consolidating, concentrating control with select transnationals. In 2009 over half (53 per cent) of commercial sales were controlled by the top three corporations, with Monsanto – the transnational with the largest market share – accounting for 27 per cent (ETC Group 2011) (see Table 3.1). In comparison, in 1996 the top ten corporations accounted for only 37 per cent of the commercial market, and Monsanto did not even make the list (ETC Group 2007a). Moreover, proprietary seed – those seeds protected by intellectual property rights (IPR) – makes up approximately 82 per cent of the commercial seed market, with Monsanto alone controlling 23 per cent (ETC Group 2008).

Through concentration transnationals build assets and eliminate licensing costs – acquiring materials and IPR of subsumed companies – while increasing their market share, increasing abilities to determine prices. For instance, after adding to their portfolio with DeRuiter Seeds, Monsanto announced seed price increases, with some corn prices rising by up to 35 per cent (OCM 2008). Taking a longer view, in the US a bushel of soybean seed cost USD7.34 in 1975 increasing to USD12.21 in 1995, but one year after introduction of GE herbicide-resistant soybeans the price jumped to USD17.40, and by 2003 reached USD24.20 (Mascarenhas and Busch 2006: 130). Even other transnationals express concern about concentration's effects. In submissions to US courts DuPont/PioneerHi-Bred (2010: 6) argues that, '[b]locking generic competition keeps prices artificially high, curbs choices in the marketplace', and that market domination – such as that exhibited by Monsanto – unfairly restricts material and information access.

---

1   The International Seed Federation, an international seed industry association, estimated the commercial seed market to be worth USD30 billion in 2005 and 45 billion in 2011 (ISF 2006, 2012), while the Context Network (2010, 2011) valued it at USD14 billion in 2001 and 37 billion in 2011. The data in-text offers a conservative, though not the most conservative, view as compromise. Despite variances among estimates, the trend of increasing value of the commercial seed market is consistent.

Table 3.1 Top 10 Commercial Seed Corporations

Corporation	Headquarters Country	Seed Sales (USD Million)	Marketshare (%, Rounded)
Monsanto*	USA	7,297	27
DuPont (Pioneer)*	USA	4,641	17
Syngenta*	Switzerland	2,564	9
Groupe Limagrain	France	1,252	5
Land O'Lakes/ Winfield Solutions	USA	1,100	4
KWS AG	Germany	997	4
Bayer CropScience*	Germany	700	3
Dow AgroSciences*	USA	635	2
Sakata	Japan	491	2
DLF-Trifolium A/S	Denmark	385	1
Top 10 Total		20,062	73

*Also listed within the top 10 agri-chemical companies

Source: ETC Group 2011

Corporate concentration not only affects commercial seed prices, but variety availability. When transnationals buy smaller seed companies, efforts to streamline production result in catalogues offering seeds with global markets and high profitability. In Canada and the US the 1980–90s buy-outs by transnationals led to the loss of regionally-adapted, open-pollinated varieties accompanied by promotion of hybridised, patented varieties. Between 1984 and 1987 in North America 54 companies offering non-hybrid varieties were bought or went out of business, and although 39 new companies began in the same period, only 21 per cent of the varieties continued to be offered (SSE 2004). More recently, larger transnationals tend to purchase smaller ones, but the effect on variety diversity is the same. The year after Monsanto purchased Seminis, for instance, the catalogue was reduced by 60 per cent, concentrating on the '25 most profitable and most [economically] sustainable crops' (Monsanto 2006a). Expansion of corporate control of seed carries with it shrinking availability of varieties; in other words, corporate concentration breeds catalogue concentration (Fowler and Mooney 1990: 124).

Seeds offer profit-making opportunities not only as commodities themselves, but also by facilitating further purchases of agricultural inputs – like chemicals. Monsanto's first GE herbicide-resistant seeds provide a case in point. Facing the impending patent expiry for their glyphosate-based herbicide, Monsanto genetically engineered seeds to be glyphosate-tolerant using bacteria genes, patented the seeds (and/or the related genetic sequences), and then sold the seeds packaged with their herbicide (Glover 2008). Profit-making continued despite patent expiry with sales of GE seed plus herbicide, and fewer costs were incurred (GE seed development requires fewer resources compared to new chemicals). Acquisitions in the 1980–90s show in these bundled commodities, and in dual market dominance – five of the six top agri-chemical companies are also top ten seed corporations (ETC Group 2011) (see Table 3.1). Monsanto's net annual income increased 104 per cent 2007–2008, based in large part on proprietary GE seed, often sold as input packages (Monsanto 2008).

Not surprisingly, given the dual concentration and profit-making potential, the dominant traits of GE crops remain herbicide tolerance and insect resistance (James 2011; Worldwatch Institute 2012), rather than, for example, improved nutrient-content. In fact, just one herbicide tolerant crop – soybean – occupies 47 per cent of the global biotech area (James 2011). The US grows the most acreage of GE crops, accounting for about 67 per cent of the global area in 2010 (35 million acres), while Canada accounts for the fifth largest acreage (8.8 million). The market value of GE crops reached USD13 billion in 2011 – approximately 35 per cent of the global commercial market (James 2011). In Canada, the Canadian Biotechnology Action Network estimates that up to 70 per cent of processed food contain GE ingredients (Sharratt 2007).[2] Only 30 GE crops were commercially cultivated in 2008, but predictions suggest that over 120 will be marketed by 2015 (Stein and Rodríguez-Cerezo 2010), some of which include patented 'climate-ready' traits such as resistance to drought, flooding or higher temperatures (ETC Group 2008; Weis 2007).

In addition to expanding commercial markets and controlling choices and prices, transnationals employ legal methods to further enclose seed saving. Corporations may layer together IPRs (discussed in Chapter 4) with other legal strategies – particularly contracts. For instance, in buying bagged Seminis seeds, growers confront a licensing agreement on their bag that reads in part: 'by opening this container you agree: (a) not to save any seeds, plants, plant parts, genetic material, parental line seed or plants or plant parts which may be found herein, and resulting produce.' Just by opening the bag, growers become legally bound not to save seed and to prevent others from doing so. Regardless of any IPRs related to the seed, this bag license constitutes a legal contract.

---

2   Sharratt clarifies that without mandatory labelling, this estimate is based on amounts of soy, canola and corn (the main GE food crops) in processed foods; GE and non-GE crops mix during processing, so assuming a processed food is not certified organic, then part of it is likely from GE seed.

Gardeners do not generally face licenses of this kind, but many seed catalogues and packets display codes for proprietary varieties, and use conditions generally prohibit reproduction. Technology Use Agreements – a more thorough version of the bag license used for large-scale growers – are common in commercial production, and stipulate production methods such as particular chemical use and timing, purchasers for resulting product, and prohibitions on saving practices (BASF 2007; Monsanto 2006b). Legal contracts along with IPRs, therefore, further enclose seed and saver options while expanding privatisation and industry profits. As consumers of seed and potential savers, growers encounter legal prohibitions on saving in multiple ways.

So far this overview of corporate seed ordering has dealt with more structural concerns, but increasing commercial market share is not the only way in which corporations advance their interests. Clapp and Fuchs (2009) explain three such methods: first, instrumental strategies – such as lobbying – directly affect other actors' decisions; second, structural strategies involve controlling others' choices by, for instance, increasing market share or privatising resources; and third, discursive strategies work to frame social norms and ideas, convincing others of the truth and normalcy of corporate preferred options. Participating in seed (and agrifood) governance allows corporations opportunities to frame, form and implement particular policy agendas, though they are not always of one mind or successful in their endeavours.

Several examples illustrate these strategies in relation to seeds. Ag-biotech representatives were fundamental to the development of TRIPs, serve as consultants at the FAO, and are well placed in the US Department of Agriculture (ETC Group 2012; Mattera 2004; Shiva 1997: 81–2). The Sygenta, Gates and Rockefeller Foundations are each affiliated with ag-biotech, as well as being involved in genebanking and promoting changes to small-scale agricultural production in the global South – including hybrid and GE seed use (Conway 2000; Syngenta 2012; Vidal 2010). In relation to influencing public perception, recently Monsanto and DuPont invested USD7 and 5 million respectively of the 34 million campaign to prevent GE labelling in California, which dwarfed the 4 million budget for promoting adoption (Cornucopia Institute 2012).

In addition to strategies used to negotiate, organise and maintain conducive technopolitical environments, corporate reordering of seed also involves and relies upon furthering seed commodification. Private capital builds on existing hybridisation practices, plant breeders' rights, and seed quality regulations, now pursuing stronger measures of neoliberalising seeds. Heynen and Robbins (2005) explain neoliberalisation of natures as an ongoing, always incomplete process involving altering governance as well as privatising, enclosing and revaluing natures. In this view, neoliberalisation includes new institutional political arrangements, shifts in relations with nonhumans toward private capital ownership and appropriations, and reduction of complex nonhumans to mere commodities. In line with this, I argue that in the contemporary seed order, as corporations – facilitated by governments – succeed in holding together and advancing neoliberalisation, seeds are transformed, becoming distanced and isolated from their historical relations, privately owned, and revalued as consumables. Seeds are, therefore, refigured. And so are growers – becoming seed consumers and business managers. Neoliberalisation of seeds is pursued through a range of technopolitical methods, and in the next three sections I consider illustrative cases, the first of which addresses seed certification.

Reusing Regulations and Revaluing Seed

At the request of government and with partial public funding, industry embarked upon a review of the Canadian regulatory environment in 2003. A year later, the Canadian Seed Alliance – a consortium of seed industry organisations including the Canadian Seed Growers' Association (CSGA), Canadian Seed Trade Association (CSTA), Canadian Seed Institute, and Grain Growers of Canada – released their report, known as the Seed Sector Review. To address the overarching aim of fostering industry competitiveness, the Review set four objectives: i) achieve regulatory flexibility and timeliness; ii) develop a more supportive and collaborative environment for science and innovation; iii) increase industry profitability; and iv) build consumer acceptance and confidence in innovation (especially biotechnology).

In the Review, industry argued that to meet these goals, major regulatory change was required – particularly regarding variety registration, intellectual property rights, identity-preservation systems and policy cooperation. In addition, the Review insisted upon continued use of risk-based assessments of GE seeds, and harmonisation of such approvals with major trading partners. Finally, it recommended a collaborative campaign with government to educate seed and food consumers about industry innovation and to create 'a level playing field between certified, farm-saved, and common seed' (CSA 2004: 2). Each of these endorsements demonstrates neoliberalisation strategies.

Most of the Review's recommendations have been accomplished or are underway. The Canadian government continues 'modernising' the Seeds Act and Regulations in line with industry suggestions for more flexible and efficient approvals – requiring less documentation, time, and merit criteria for new registrations, with decisions being led by consultative groups (in which industry representatives play significant roles). Further, in 2011 the federal government initiated a Seeds Value Chain Roundtable (SVCR) for policy-making collaboration with industry, continuing the work of the National Forum on Seed that served a similar purpose 2005–2009, until government funding ceased. Moreover, identity-preservation systems are increasingly used, public research progressively deals with base data rather than commercial variety release, the Canadian Wheat Board has been massively revamped, contract growing is increasingly popular, and the 'substantive equivalence' risk-based approval of GE seeds remains. On the whole, the Review's recommendations successfully found their way into reworked governmental policies.

Industry continues to push their successes further, while also working to attain those goals unachieved. In the new SVCR, for instance, industry already argues for more crops to be moved to lower assessment levels for registration, articulates concerns about deregistration of heritage varieties, and questions some uses of variety names (Kozub 2012). Strategies to strengthen IPR and increase certified seed use have been less successful, but both industry and government continue to advocate adoption of UPOV91 (see Chapter 4), and industry has initiated a certified seed campaign. It is this last endeavour – promoting certified seed – that I investigate next.

Certifying Seed

Certified seed must be a registered variety of seed produced under specifications to meet standards for purity, germination and disease. It is the final outcome of pedigreed seed growing that multiplies breeder seed for commercial marketing, and it is bought and sown for crops.[3] 'Certified' is one of the seed classes laid out in Canada's *Seeds Act*, the first form of which set basic standards for seed quality in 1905. Initially enacted to protect growers from dishonest seed sellers, the Act now governs commercial registrations, sales and imports, inspections, and quality assessments of specified crop varieties. In recent years government, on industry advice to 'modernise', has reworked the Act and its regulations. Though the pedigreed seed process and its classes remain unaltered, industry now turns to its advantage provisions once used to protect growers from profiteers. In promoting certified seed, industry does more than advocate quality seed – it attempts to enclose seed saving practices, expand certified seed markets and increase opportunities for profits.

Seed becoming certified involves a complicated process implicating growers, a regulatory framework, industry associations, standards, documentation and a governmental agency. To begin, growers obtain pedigreed seed for growing out – on contract or by purchasing it from corporations or growers. Parameters must be met regarding previous land use, minimum isolation distances and crop purity. As crops are growing, any off-type plants – rogues – must be removed. Additional contract terms – like which pesticides to use and when, or crop-growing requirements such as growing hybrids in rows of 'females' and males – fall outside certification requirements, but influence certified seed growing.

---

3   The process for all pedigreed seed classes is similar, but higher classes must meet more stringent standards.

Seed growers submit applications and fees to the CSGA to have their seeds certified. The CSGA was originally part of the federal government but since 1926 has operated as a separate association with a board of industry and appointed government directors, and now represents approximately 3,000 seed growers (CSGA 2011). The CSGA checks applications and sends inspection requests to the Canadian Food Inspection Agency (CFIA). As the crop is growing, the CFIA's governmental fieldstaff perform inspections. In 2010 the CFIA announced it would withdraw from inspection within five years, and accredited associations or companies would be taking over (Dawson 2010a), but whether and how this will occur is unclear. After inspection, reports are returned to the CSGA, which then assesses the results and, assuming standards are met, issues a crop certificate.

As growers harvest the seed, field samples are taken for germination and purity testing. Private labs and processing facilities – accredited and monitored by the industry-led Canadian Seed Institute – complement the random sampling done by the CFIA. If seed meets the industry-set standards, it receives an official certified seed tag and document – which enable the grower to label and sell the seed as certified and by its variety name. In order for certified seed to retain its designation, accredited processing and grading companies must also meet industry-specified standards. Enforcing compliance is the CFIA's responsibility. It takes, in sum, a collaborative and systematic effort to produce certified seed.

Attempting Revaluation

The Seed Sector Review identifies certified seed as undervalued, arguing that market expansion is necessary to increase industry profitability. Recommended measures include: increasing certified seed use; educating growers of the benefits of certified seed; revaluing saved seed by publicising it as lower value and by raising its costs; and restructuring seed systems (CSA 2004; Fehr 2005). Since the Review, industry has succeeded in having some crop insurance and loan rates adjusted when certified seed is used, as well as having certified seed increasingly specified as a minimum requirement for identity-preservation and other production contracts. However, other Review recommendations for mandating purity standards for saved seed, royalty collection on saved seed, and a tax incentive[4] for certified seed remain elusive.

Industry associations and corporations persist in efforts to revalue certified and saved seed. The CSTA and CSGA collaborate in running a multi-faceted certified seed campaign directed at domestic and international seed and food consumers including videos, mailings, website development, presentations, and

---

4   Recognising cost as a limitation to certified seed use, industry lobbies for Canadian income tax law revision to allow growers using certified seed to claim 155 per cent of its cost as part of expenses (CSTA 2011b).

so forth (CSGA 2011). The CSGA, for instance, informs its members who sell certified seed that '[i]t is easy to promote one variety over another but selling seed also means explaining the benefits of certified seed to farmers and the value that it can provide compared to farm saved seed' (CSGA 2010: 18). In promotional literature, certified seed is advocated as advantaging growers with early access to new varieties, high purity and germination levels, and traceability (Adolphe 2011; CSGA nd). One campaign document explaining *The Value of Certified Seed*, asks growers: '[i]n an era of herbicide tolerant crops and specialty traits, can you afford to get anything less than the highest possible varietal purity?' (CSTA nd). Industry's goal of increasing certified seed use is understandable given the profits that would come from new purchases, but would it serve growers?

Unfortunately for industry, saved seed is not necessarily substandard compared with certified seed. As long as growers take appropriate measures in selecting, cleaning and storing seeds, saved seed equals the quality of certified seed. As plant scientist Steven Shirtliffe (2007) explains, '[t]here is nothing "magic" about certified seed – it is simply seed that has been grown under careful and documented conditions'. Pedigreed seed growers may tend to be more careful in their handling than other growers particularly when it comes to mixing seed – the most common cause of impurity – but keeping seed weed-free with low disease levels is usually sufficient for growers' needs (Shirtliffe 2007). In this sense, certified seed may be unnecessary to meet growers' needs for quality seed.

Farmers seem to agree that saved seed is at least as good as certified seed. Paul Beingessner, a Saskatchewan farmer, states, '[t]here's lots of seed trading among farmers here. We rarely buy certified seed for cereals. It's rarely better seed and just not necessary' (quoted in Leahy 2005). Farmers' continued belief that there is no major quality difference between certified seed and saved seed was confirmed by results of focus groups held by the Canadian Wheat Board in 2005 (SeedQuest 2005) and by industry surveys revealing that a substantial majority of growers prefer to use saved seed for major field crops, except canola (Allen 2007; CSGA 2005; Blacksheep Strategy Inc 2004). For instance, saved seed accounts for over 80 per cent of wheat, barley and pea production, and about 75 per cent of flax (Dyck et al. 2010; LeBuanec 2005). Growers may replenish their saved seed with small amounts of certified seed every 3–5 years, or, to achieve similar ends, exchange among themselves. The area seeded with certified canola seed is high (80–90 per cent) in Canada for various reasons including the widespread use of hybrid seed, contractual obligations for those using Roundup Ready and Clearfield varieties, and on-farm seed treatment limitations (Holm 2007). In many ways it is these conditions that industry continues pursuing for other crops.

Saved seeds offer not only comparable quality – based on purity, germination and disease standards – but also further benefits such as lower costs and local adaptation. An estimated CAD47 million would be saved by Saskatchewan wheat growers sowing 90 per cent of wheat acreage with saved seed (Beingessner 2004a). Addressing a Canadian Senate Committee discussing the future of agriculture, Neal Hardy, a farmer and director of the Saskatchewan Association of Rural Municipalities at the time, explained:

> Our concern is that if changes to the pedigree seed sector require farmers to use pedigreed seed and to buy it from a seed distributor rather than using their own seed, costs will go up dramatically. We all buy certified seed at times to re-establish our seed, but many of us carry our seed over. I personally do that for a couple reasons. One is that it is much more cost effective and I have high a [sic] germination rate. Second, it is acclimatized to my area. I am from the northern part of the province and it shortens my season. Some will say it does not, but it does. If I buy my seed from the south, my season will be five or six days longer than if the seed is from my own area. (as in SSCAF 2004)

To Hardy's mind, saved seed meets his needs better than certified seed by combining high quality (germination) with cost effectiveness and seasonal adaptation. Accordingly, industry's push toward certified seed presents a problem rather than a solution for many growers.

If growers are to buy certified seed, then it makes sense to source seed from within their region – given adaptations to climate and regional economic benefits. However, not all varieties are available regionally or even nationally. Practices of growing pedigreed seed are declining in Canada. Pedigreed crop acreage has dropped significantly since 2001, in 2010 settling at just over 1.1 million acres dominated by wheat, soybeans and barley (Adolphe 2011). Contracted growers are often required to use specified varieties and sources – as noted, something industry supports – but specified seed is not necessarily Canadian-grown. For instance, canola is not high on the list of pedigreed seed grown in Canada despite its high production acreage and certified seed percentage. As seed and growers become further enrolled in transnational ordering, use of imported seed is likely to grow.

While pedigreed seed growers work hard to provide quality seed, and certified seed does facilitate quick access to new varieties, it is not enough. Despite all its purported benefits, certified seed does not necessarily meet the needs of growers. In summarising the industry-commissioned survey in their publication, the Canadian Seed Growers' Association (2005: 4) discloses that:

> [f]ully half or 50 per cent of farmers think weed-free common [saved] seed is equal to Certified seed. ... And here's another consideration: Are farmers purchasing Certified seed because they think it offers great value for the money? No. Almost 20 per cent of those purchasing Certified seed report they did so because common seed was not available.

In this context it seems reasonable to at least consider the option of increasing, not decreasing, practices of saving seed. Convincing or cajoling growers into 'choosing' certified seed is ill-advised given its overall benefits and growers' preferences. Unfortunately, possibilities of increasing saved seed are not high on the priorities of industry-government collaborations. Instead, focus falls on fostering 'innovation and a friendly regulatory system' for industry (Febr 2005: 8).

Much neoliberalisation involves amending regulation – as with seed registrations – but long-standing policies such as certification processes may also be enrolled in new ways to create and expand markets. As this section illustrates, certifying seed with industry reinforces neoliberalisation processes. First, industry promotes seed enclosure by further distancing growers from seed saving, urging growers to buy seed as well as requiring seed growers to use higher class seed and possess officially recognised expertise to qualify as certified seed growers. Second, the promotion of market-based incentives and minimum contract terms not only expands certified seed markets, but serves corporate interests of privatisation – making seed and growers beholden to corporations that increasingly dominate research, distribution and marketing. Third, certified seed revaluation strategies not only work to raise some seeds above others in questionable ways, but continue processes through which seeds are reduced to commodities, and growers to consumers. In the next section, I turn my attention to other ways in which neoliberalisation proceeds by examining attempts to alter policies in connection with GE contamination of non-GE crops.

Contaminating Seed and its Remobilisation

Despite the rapid spread of GE seeds, their introduction and use continues to be controversial. On the one hand, GE proponents – some more cautious than others – argue that GE provides opportunities for more choice, increased production and environmental sustainability (Brooks and Barfoot 2011; FAO 2004a, 2010a; Paarlberg 2008). In this view, GE is said to provide the latest innovations inserted into the best germplasm, with ag-biotech serving to improve lives (see Dow AgroSciences 2007; Gates Foundation 2012; Monsanto 2012). Critical engagements with GE, on the other hand, discuss the risks and consequences for humans and nonhumans, limited scientific and ethical consideration, problematic political-economic arrangements, and dispute claims that GE is necessary for food security or environmental sustainability (see Altieri and Rosset 2002; Bellon and Berthaud 2006; Benbrook 2012; Buttel 2005; FOEI 2011; Jasanoff 2005; Kloppenburg 2004; Kneen 1999; McAfee 2003, 2008). Further, contestations of GE foods are evident in publics and policies, at national and international scales – manifesting (among other things) as uprooted GE fields and mandatory labelling demands, WTO disputes and Cartegena Protocol enactment.

These contentions and contestations warrant serious critical attention. But my focus here falls on neoliberalisation with corporations, and in the next section I investigate a particular event, flax contamination, which illustrates the relations of GE seeds, growers, industry, international trade, market valuations and re/regulation initiatives.

Contaminating Flax

Canadian-grown flaxseed was found to be contaminated by GE in 2009. Flax is used for animal feed and industrial purposes, but also as food – marketed as a healthy grain, with high omega-3 fatty acids, fibre and protein levels. Even before GE presence in the non-GE flax was confirmed, prices for flax almost halved and its trade was suspended between Canada and Europe (SFDC 2009). The report of contamination, initiated by the European Union's Rapid Alert System for Food and Feed, was joined in following months by contamination detections in 35 other countries around the world (CBAN 2012). In importing countries products were recalled, processing halted, customers refunded, and flaxseed products increased in price due to limited uncontaminated supplies (COCERAL/FEDIOL 2010). Much of the Canadian flaxseed ready for export sat in storage, prompting the Saskatchewan Flax Development Commission President to comment that 'the flax market has basically collapsed' (Kuhlmann 2009). Cleanup programs initiated multisite testing throughout the production process, with destruction or redirection of contaminated crops. Depending on the market, product use, and approval status, different standards and responses for GE contamination apply, but contamination is always a threat for non-GE growers and consumers.

Before the contamination, Canadian-grown flax accounted for 44 per cent of world production (FAOStat 2009) with around 75 per cent of exports going to European countries (CGC 2008, 2009). Contamination changed things significantly. In 2010/11 Canada's flaxseed trade with Europe was less than half its 2007/8 level (CGC 2008, CGC 2011), and the area in production was 45 per cent of that in 2009 (Statistics Canada nd). Further, the contamination prompted Europe to seek other suppliers and damaged the reputation of Canadian-grown flax, which combined may prove more damaging long-term than the direct effects. Production and exports have yet to recover despite a new protocol, which allows contamination of 0.1 per cent for feed and .01 per cent for food. But prices are on the rise, production area is projected to increase, and seed testing indicates much lower levels of contamination as cleanup efforts continue (FCC 2011, 2012).

Cleanup protocols consider GE regulations of destination markets, but also seed materiality and agrifood chain processes. The flax plant is self-pollinating and no GE flax was authorised for production, so the source of contamination was expected to be seed – rather than gene transfer. In addition, flaxseed is quite small and – whether wet or dry – it sticks to equipment, making not just seed supplies but handling and transportation processes possible contamination sources. Initial thoughts that the contaminant was residue from other GE crops – canola or soybean – were dismissed when GE flaxseed was detected.

GE flaxseed is not approved in Europe, nor is it registered for commercial production in Canada. The Crop Development Corporation at the University of Saskatchewan developed a GE flax (CDC-Triffid), and the CFIA and Health Canada approved it for animal and human consumption in the late-1990s. But because primary export markets for Canadian-grown flaxseed do not accept GE flax, growers concerned about losing markets due to contamination mobilised against the variety's release. Unable to have safety approval rescinded, flax producers, the Flax Council of Canada (FCC), and the Saskatchewan Flax Development Commission (SFDC) lobbied for variety deregistration – which would make it illegal to sell or commercially grow the GE flax despite its safety approval. In 2001 the variety was deregistered.

Deregistration often occurs as varieties become commercially unprofitable, but this was the first time deregistration was sought and obtained at growers' insistence. At the time, pedigreed seed growers were multiplying the flax for commercialisation, but with deregistration those crops were destroyed and the CDC indicated they would stop exploring GE flax options (Warick 2001). For Canadian flax growers losing markets to GE flax contamination eight years after taking measures to prevent just such an occurrence was a shock.

GE contaminated flaxseed reveals a unique, but familiar story. It is not the first – or even most recent – detection of GE contamination.[5] GE contamination has been found in wild and cultivated organisms, in food, feed, and industrial chains, and has spread through human error, unfamiliarity, and misrepresentation as well as organisms' reproduction. It is also not the first time GE contamination resulted in significant social and economic costs – for producers, industry, governments or consumers. Given the spread and continued development of GE crops, and continuing public concerns about food safety and environmental impacts it is unlikely contamination events or their impacts will abate any time soon. They are expected, and it is becoming a major trade issue.

Remobilising Contamination

Efforts to regain European flax markets continue in Canada, but this event's reach extends beyond flax exports. Both ag-biotech advocates and those more cautious of its impacts – real and potential – attempt to use this event in advancing their agendas. In the aftermath of the event, industry strategies

---

5   For a selection of GE contamination cases relating to seed see Amagasa (2007); Bondera and Query (2006); Clapp (2005); Hall et al. (2000); Marvier and Van Acker (2005); Miller et al. (2006); Quist and Chapela (2001); Warwick and Meziani (2002); Zerbe (2004).

focus on expanding markets through protocols for certified seed and regulation for low-level presence (LLP) policies, while others attempt to prevent market loss through proposals for revising policies and approval processes, and/or establishing compensation mechanisms. Each attempt brings controversy.

Enlisting Certified Seed

Certified seed became part of the GE flax contamination story through the cleanup protocol. In the lead-up to the 2010 season, the FCC announced a requirement for farmers to use certified seed instead of saved seed, advocating it as a solution to the ongoing contamination problem (Dawson 2010b; FCC 2010a). Three quarters of flaxseed is grown from saved seed (Dyck et al. 2010), and normally flax growers buy small amounts of certified seed every few years to supplement their saved seed. Implementation of the FCC's requirement would increase costs and sacrifice locally-adapted seed, and since the source of contamination was unconfirmed at the time the requirement seemed unreasonable to some. The SFDC and the National Farmers Union (NFU) argued against the measure, insisting that certified seed is not necessarily more pure than saved seed and that all seed supplies – saved and pedigreed – should be tested as part of measures to restore markets (Dawson 2010b; NFU 2010; Pratt 2010a).

Despite reassurances from the FCC that the move was just about regaining European markets, some growers and farmers' organisations expressed concern that this was really about stopping farmers from saving seed and establishing certified seed markets (as previously discussed). Terry Boehm, NFU President and flax grower, warned: 'We cannot let companies act opportunistically to leverage this flax sector problem into an opportunity to boost the price of flax seed and force farmers to buy all their seed' (as in NFU 2010). Similarly, Allen Kuhlmann, President of the SFDC, explained the difference in association positions by stating, 'I guess we come from different places. The Flax Council's budget comes from the people who handle flax – Viterra, Pioneer, Cargill, and so on. Our money comes from producers' (as in Pratt 2010a). In the face of this resistance, pedigreed seed was tested. When several seed stocks were revealed to be contaminated, the FCC revised its stance arguing that '[t]he best option for producers remains the planting of certified seed', but that saved seed could be used provided it passed the same testing protocols as assessed by industry-approved labs (FCC 2010b).

Confirmation that pedigreed seed served as the source of contamination meant that cleanup efforts needed to alter; the testing in fields and key points in the seed chain were necessary – as contaminated seed might have endured – but insufficient. The contaminated lines have since been removed from the market, and breeding efforts to reconstitute clean seed stocks have begun. So far, though it continues its advocacy, industry has been unable to justify mandates for certified seed as contamination cleanup.

Advocating Low-Level Presence

The flax contamination and economic losses incurred serves also as one of several cases that industry associations employ to argue for acceptance of LLP policies. Industry, and increasingly governments, use the term 'low-level presence'[6] when referring to contamination by GE crops that are unapproved where contamination is found but that are approved in at least one other country. CDC-Triffid flax provides an exemplar: it was unapproved in Europe, and approved – though not registered – in Canada. With a LLP policy, a certain amount of GE would be allowed in food and feed, regardless of approval. So, depending on the established threshold, a policy of this kind would have made the GE flax contamination moot and exports and processing would have continued.

For ag-biotech advocates, LLP policies would deal with multiple problems. According to industry associations, the rejection of unapproved GE results in trade and business disruptions and prevents exporting countries from adopting new technologies that increase productivity and competitiveness (CropLife International 2012a; CSTA 2011a). 'Zero-tolerance' policies are not an option according to ag-biotech; instead, LLP acceptance is necessary. In addition to thresholds to allow unapproved GE, ag-biotech endorses LLP policies that integrate industry self-regulation, global GE approvals and scientific assessments of relative risk for GE.

As the CSTA (2011b: 1) outlined in a presentation to the Canadian government, the private sector – in addition to its own promotion of LLP – seeks from government:

> A commitment to using all possible negotiating and other measures to facilitate the international trade of seed, specifically the development and implementation of an international policy on low-level presence of genetically modified products which have been approved as safe for food, feed, and environmental release in countries that employ sound science, but not in the country of import. (CSTA 2011b: 1)

Several points made in this quote are note-worthy, including: industry's demand that all possible measures be used; the unquestioned pursuit of international trade; the clear distinction between countries approving GE with 'sound science' and others; and an implied inevitable, but unproblematic, presence

---

6   Adventitious presence, once employed by ag-biotech advocates for all contamination events, is now used more specifically when GE contaminants are unapproved in any country – such as in recent detections of Bayer's GE LibertyLink rice (Gillam 2010a, Harris and Beasley 2011). With either term's use, contamination does not appear – instead, language choices reinforce a portrayal of GE presence – even if illegal – as benign and accidental.

of GE. The Seed Association of the Americas, CropLife International, and the International Seed Federation – all international industry associations – make similar demands of state governments and multilateral bodies.

In support of industry, and in many cases using its language, the Canadian government pushes for a global approach to LLP within the Organisation for Economic Co-operation and Development, the Codex Alimentarius, and the WTO, as well as within bilateral agreements. Agriculture and Agri-food Canada is also reconsidering domestic policies. Canada's advocacies for LLP acceptance include endorsement of (among other things): synchronous GE approvals; minimised trade disruption; risk assessment and mitigation; and facilitation of agricultural innovation (AAFC nd). The recommendation for global acceptance of approvals would require regulatory change in Canada, as would further moves toward industry self-regulation. But the general approach, and even the specifics of science-based risk assessment, is consistent with North American approaches.[7] The AAFC's initial suggestion for these policies to permit unapproved GE levels in food up to 0.1 per cent is controversial. Higher thresholds are advocated by industry, questions posed by various interests around different state safety approvals, and resistance to LLP by some states, non-governmental organisations and organic producers remains (Bruer 2012; CBAN 2012; Schmidt 2012).

Proposing Economic Assessments

Arguments for LLP illustrate savvy redeployment of a contamination event to assure and stretch markets through reregulation, but the same case justifies manoeuvres to retain non-GE markets and production. For instance, not long after the flax contamination, Bill C474 was introduced to Parliament as a private member's bill proposing Canada's GE approval process be revised to include economic impact criteria. Responding to the proposed Bill, the General Manager of the Canadian Canola Growers Association argued that, with the Bill, the canola industry might lose the benefits gained from GE canola, which now dominates the sector, and that further research and development would be impacted. Instead, he endorsed a LLP strategy, adding, '[w]e strongly encourage Canada to stick to our guns on science based regulatory processes. Keep the politics out of it' (as in Pratt 2010b). Dow AgroScience's representative agreed, contending that regulating LLP for unapproved GE crops was the best course of action (Pratt 2010c), and Monsanto's spokesperson warned against politicising GE approvals by allowing comments from those uninvested in industry successes (Dawson 2010b).

---

7   For various critiques of Canada's approach to GE regulation see Abergel (2000); Andrée (2002); Bocking (2000); Clark (2003, 2005); Kneen (1999); Mauro et al. (2009); Office of the Auditor General of Canada (2004); Royal Society of Canada (2001).

In contrast to ag-biotech advocates, the Executive Director of the Canadian Organic Trade Association (COTA) and the President of the NFU detailed how GE crops are not desired by many markets and how, therefore, Canadian exports are threatened by GE contamination, making economic assessment a reasonable requirement (Ewins 2010; Wilson 2010). Further, COTA argued that if existing production and sales are not to be considered, then policies need to assure ag-biotech provides compensation when markets are lost to contamination (Wilson 2010). The Bill did not pass; however, questions about market acceptance, contamination compensation and GE approval processes persist. Debates continue over the researching and releasing of GE versions of alfalfa, sugar beets and mustard, as well as the return of GE wheat – already once repelled by Canadian growers.[8]

Market Opportunities
In this case of GE contamination of flax, market logic dominates both sets of responses – one arguing for regulatory harmonisation to facilitate trade, and the other arguing to maintain economic advantage. LLP continues work to normalise GE crops through regulations and pervasive presence. Agbiotech has, since the 1980s, pursued global GE acceptance and scientific risk assessment approval policies (Andrée 2007). LLP policies would open markets in new territories – like GE-resistant Europe – as well as limit alternative market gains, offering means of furthering profit-making, control and neoliberalisation. Moreover, the LLP strategy defines the problem as one of regulatory asynchronicity rather than fundamental conflict – the solution becomes policy harmonisation rather than ethico-political engagement. However, such policies do not happen on their own, they are adopted by governments, accepted by publics and promoted by corporations. Keeping 'politics out of it' is part of neoliberalisation processes, one effect of which is to 'sequester key economic policy issues beyond the reach of explicit politicization' (Tickell and Peck 2003: 175), serving to deny what is a political process that serves particular interests.

GE contamination presents a marketability problem for non-GE growers, but their responses differ from those of ag-biotech. While ag-biotech endorses alleviating 'over-regulation' through international acceptance of GE and risk-based assessment criteria, critics endorse more preventative measures – working to prevent commercial release, but otherwise advocating for market assessments and compensation, if not GE-free zones (*cf.* Altieri 2005; CropLife International 2012a, b). The economic leverage of growers and consumers provides opportunity to question particular forms of market logic being used, and to push connected ethico-political concerns. The NFU (2000), for instance,

---

8   In 2004 public pressure compelled withdrawal of Monsanto's GE wheat application in Canada. Research relaunched in 2009 and field trials are underway in Australia (Cubby 2011; Gillam 2010b; Monsanto 2009).

connects market loss and compensation issues arising from GE contamination with corporate control and human and nonhuman health effects. Despite maintaining market logic and assumptions of the possibility of coexistence therefore, contestation exists here.

Contamination and LLP proposals do not directly promote or impede seed saving practices, but their relations bring difficulties. As revealed, one response to contamination was promotion of certified seed – even, at one point, mandatory stipulations that were later revised. Increased certified seed use, as discussed in the earlier illustration, would negatively affect saving seed practices. LLP policies' implications for saving are also indirect. Proposed LLP policies' spread of GE would come with increased presence of IPRs and, most likely, corporate licensed growing practices, each of which shut down opportunities for legal saving. Even the fear of IPR infringement may be enough to dissuade growers from saving seed, despite its many benefits. Though unlikely, the controversy could also raise resistance or just a desire to control seed sourcing, and increase seed saving.

In sum then, skillful integration and remarketing of problems by ag-biotech advocates turns contamination, through LLP and certified seed advances, into opportunity. Corporate reach stretches and neoliberalisation deepens as GE acceptance is legislatively mandated and made material, though not without contest. Recognising the ongoing contestation and incompleteness of regulatory strategies discussed so far, ag-biotech also seeks more penetrating dominance of seed through genetic engineering methods, the subject of the next section.

Engineering Commodities

Development and adoption of GE crops has been swift – with seeds moving from lab to commercial growing in just over 20 years, and increasing in acreage annually. Each GE organism is a collective enactment – one materially and practically different from other seed breeding in labs or fields. In this process, seeds – and their plants – are transformed into a different kind of socionatural artefact. Much of the difference of GE seeds in comparison to other seeds has to do with a combination of unprecedented levels of manipulation of biological processes and sizable amounts of required resources – economic, material and informational.

While the previous section illustrated the implications and debates over a case of GE presence, this section examines genetic engineering methods and the implications of their connections with corporations interested in seed reordering. In considering genetic engineering's potential contribution to deepening seed commodification and participation in neoliberalisation, I focus briefly here on one extreme example of engineering genetics – genetic use restriction technologies (GURTs) or Terminator seeds, but first I take a brief look at the breeding process involved.

Engineering Genetics

The practices of genetic engineering allow the marking, removal and manipulation of organisms' genomes as well as the transference of genes between species that would not normally interbreed. Methods of genetic engineering refine and develop, raising new possibilities and concerns, but most commercialised GE seeds participated in a process similar to the one described below.

To breed GE seeds, lab-based scientists extract an identified gene or genetic sequence from an organism – bacteria, plant or animal. The extracted gene(s) is then marked (usually with antibiotic-resistant genes), perhaps connected with modulating genes to regulate expression, and recombined with a vector (usually a bacterium). This vector material is then reinserted into plant cells by bacterial infection or gene gun. Not all plant cells accept the transfer, so material is sorted – often by applying antibiotics that allow marked antibiotic-resistant sequences to continue, but kill untransformed cells.

Once sorted, transformed plant cells are regenerated using tissue culture methods. Successful regeneration, therefore, results in plantlets with transferred genes in each of their cells. The GE plantlets are tested to ensure they carry the desired gene, and in the preferred amounts. In some cases, resulting GE plants are crossbred with others that have other targeted traits. Once the GE plants reproduce seeds judged adequately stable, commercialisation ensues. This process description oversimplifies the work and relations that go into generating GE, both in the laboratory and beyond it, but gives an idea of the method. The next section connects GE methods and relations through the example of GURTs.

Terminating Possibilities

Ag-biotech developed GURTs through genetic engineering as a biological way to protect their proprietary interests. GURTs operate either at the variety level, controlling a seed's capacity to reproduce, or at the trait level, directing de/activation of particular traits. In either case, genetic sequences engineered into seeds include 'switch mechanisms' to prevent unauthorised use. I focus here primarily on the varietal restrictions (V-GURTs), though much would also apply to trait applications (T-GURTs).

V-GURTs are genetically engineered seed that may be chemically induced in one of two ways. In the first, a disrupter gene – a modulator included within the genetic sequence – is triggered with chemicals at the point of sale. In this variant, in the final development phase the GE plant produces a toxin that kills embryos and leads to the production of sterile seed. In the second kind of V-GURT, the disruptor serves as default. For this type, chemical is applied throughout seed breeding and then stopped at point of sale, allowing the disruptor to express. The seed produced, therefore, is sterile unless reapplication of the chemical occurs at the appropriate time.[9] In either case, growers who use these seeds will produce a crop but be unable to save viable seeds. For these GE seeds, reproduction occurs chiefly at the command of corporations, chemicals and licensing payments.

The first patents on GURTs sequences were claimed in the US in 1998 by Delta and PineLand (DPL) (bought by Monsanto in 2006) and the US Department of Agriculture (DPL and USA 1998). In response to GURTs development, the non-governmental organisation RAFI, now the ETC Group, released information on the technique, its patenting, and redubbed GURTs as Terminators – a name that has stuck. DPL and the USDA had planned to license rights to use Terminator widely, allowing ag-biotech transnationals to 'increase the value of proprietary seed owned by the US companies and to open up markets in Second and Third World countries' (USDA spokesperson as in Engdahl 2006). But that plan was put on hold.

Swift and intense lobbying from non-governmental organisations propelled response from the international community. Monsanto, acknowledging the risks of GURTs, pledged never to use the technology (Pendleton 2004). But this offers little assurance. In addition therefore, the Convention on Biological Diversity's Conference of the Parties signed a '*de facto* moratorium' on Terminator in 2000, urging countries to abstain from fieldtesting and commercialisation until reliable data proved the need for the technology and its safety. The agreement persists, despite repeated efforts to revoke or weaken it. Canada, for instance, signed the agreement but continues to argue that testing should proceed on a case-by-case basis. It does so despite its admission that it is unprepared to test GURTs, and has not engaged in public consultations (CBAN 2006a, b). Further, research and development continues through corporations such as Syngenta, DuPont and Monsanto, governmental organisations such as the European Union and the USDA, and academic institutions including the research foundations at Cornell and Purdue Universities. Patents are now held on various Terminators.

In the face of international resistance to so blatant a play for corporate seed control, ag-biotech continues working to rebrand Terminators with modified justifications for, and discourse about, these seeds. Ag-biotech maintains that these 'technology protection systems' protect IPRs and profits, but has added two other benefits. First, they say, GURTs promote research and development by protecting intellectual property, ensuring a competitive industry and more choices for growers.

Second, and more strategically, Terminators are promoted as biosafety measures offering protection against undesired gene and pollen flow (see Collins 2006; Collins and Krugueger 2003). A DPL (2005) pamphlet explains:

---

9   In T-GURTs the target trait is switched on or off through chemical induction of gene silencing or by separating, or removing, genes.

TPS provides the biosafety advantage of preventing even the remote possibility of transgene movement. Even in the unlikely event that a TPS plant were to pollinate flowers of a wild, related species, TPS would prevent the resulting seed from spreading the transgenic trait. TPS also provides the additional biosafety advantage of preventing 'volunteer' plants, a major problem where crop rotation is practiced.

Here DPL walks a fine line – attempting to reassure that GE contamination is a remote possibility (therefore retaining existing acceptance), while presenting Terminators as solving a problem (thereby creating a new market). Through this lens, while GURTs would not have prevented the contamination event discussed in the previous section (since it was not due to geneflow), other problematic GE presences – such as 'superweeds', plants that have bred with herbicide-resistant GE plants and acquired their traits – might be prevented.

Critics of Terminator seeds and traits dispute ag-biotech's characterisations.[10] Instead, critics contend that Terminators would increase grower and consumer dependence on corporations, reduce availability of and access to appropriate seeds, and spread the already existing problems presented by GE crops – both by expanding markets for existing GE seeds and by commercialising new ones (Boehm 2006; ETC Group 2003, 2007b; UNEP 2003, 2005a, b; Visser et al. 2001). So far, Terminators have not been commercialised. However, if governments can be convinced that GURTs serve useful purposes with limited risks this will change. Indeed, future regulators so convinced may actually require Terminator GE seeds for biosafety (ETC Group 2007b), resulting in corporations being paid to protect their own interests.

Terminator seeds embody practices and logics working for seed control through technological interference with seed reproduction. They are the clearest – and most extreme – example of ag-biotech transforming seeds into commodities, subsuming and enrolling them in corporate seed ordering. Terminators represent an unprecedented level of interference and control. As Van Dooren (2007) notes though, GURTs do not eliminate historic seed-people relations, and in this sense Terminators' reach is limited. In addition, the unpredictability of genes and seeds, so obvious with GE contaminations, would continue, even if to lesser degrees.

With Terminators the legal time and space limits of IPR and contracts become less problematic for transnationals as requirements to buy seed anew, or at least pay again, become part of seed bodies. By transforming seeds in such a fashion, ag-biotech ensures and expands markets and furthers neoliberalisation by turning a seed's capacity to embody against it. Further, the deaths of seed kills

---

10   For further considerations of possible benefits and limits of GURTs see Eaton et al. (2002); Goeschl and Swanson (2003); Hills et al. (2007); Van Acker et al. (2007); Van Dooren (2007); Visser et al. (2001).

not only their ongoing stories and lives, but also future possibilities to secure food, to adapt to changing environments and cultures, and to form relations with savers through saving. In a very real way, with Terminator, the lives and deaths of these seeds, and all the affordances they offer savers, would belong to ag-biotech. Despite international restrictions on the technology, Terminators' return is not unexpected, particularly given new biosafety justifications. They are aptly named then, not only for the death they bring but also for their reworked recurrence.

Engineering Enrolments

The increasing domination by transnationals of seed chains, in part, relies upon the opportunities of genetic engineering. GE breeding raises possibilities for new products – like seed with genes from bacteria for killing insects. Further, these products take less time to develop than other lab-bred seed (Louwaars et al. 2009). Extracting, isolating and manipulating genetics also facilitates ownership claims, making it easier to claim invention of sequences, genes and processes. In addition, GE allows isolated traits to be combined – or 'stacked' – in one organism, prolonging commercial sales of particular traits as part of renewed product lines. Moreover, GE research reveals that seeds may not only be manipulated to resist or express particular chemical formulations; instead, seeds can be transformed to require particular inputs to 'trigger' trait expression – as in Terminator technologies.

However, transnationals also condition GE breeding. Engineering genetics is both complex and resource intensive, requiring specialised expertise and equipment as well as access to material and knowledge, often already privatised. Transnationals assemble the required resources providing possibilities for new GE seeds in ways few others can manage (Carolan 2010). Even with released public varieties, cultural and material shifts to marketable research mean that more often than not private capital influences with funding, intellectual property licensing, and/or market priorities. GE seeds, at least so far, rarely make it to commercial release without transnationals' cooperation, if not control.

In short, transnationals capitalise on the possibilities of GE to expand markets and control, increasing profits while mobilising resources and reinforcing GE acceptance and relevance in particular ways. Of course, the ag-biotech industry could not reorder seed with GE without governmental supports – such as subsidies, shifts to breeding and production cultures, reregulation favouring strong IPRs, quick and easy approvals, and limited industry responsibility for contamination. By exploiting technological and political opportunities for neoliberalisation, transnationals prosper. GE seeds, for their part, live with and embody knots of technoscientific and socionatural processes, reinforcing and relying upon transnational domination of commercial seed.

The 'best possible germplasm'

> In the competitive seeds-and-traits business, our customers are interested in the complete package of a seed and its genetic make-up. The reason is simple: farmers buy yield. They want the seed that has the strongest potential for the highest yield, and they want the tools that allow them to protect that yield potential against environmental factors such as disease, insect damage and weed competition. Meeting farmers' demand means offering the best possible germplasm, incorporating the newest generation of biotech traits and combining those biotech traits into the combinations that allow farmers to do more with every seed. (Monsanto 2007: *par.* 4–5; see also 2005b: 5)

Monsanto defines its purpose here as providing 'the best possible germplasm' to meet growers' demands. Growers, in this view, do not buy seeds; seeds are a delivery mechanism – a convenient container – for 'the newest generation of biotech traits' that comes with a complete growing package for consumers. Monsanto, according to this quote, promotes itself as really being in the business of providing choices – 'tools' for their grower-customers.

The idea that new commodity-technologies provide more and better choices is fundamental to the success of private capital and sits easily with prevailing consumer culture. Industry argues that commercial, especially GE, seeds are adopted because they offer growers a better choice. Technological innovation for the market, goes the argument, enables liberation while leaving open each individual's pursuit of the good life (see Borgmann 1984; Mesthene 1983).

But seeds – GE or otherwise – do not exist in isolation as neutral tools; rather, they are the result of, and immersed within, technopolitical relations that facilitate and condition particular realities. It is necessary, therefore, to examine the ways in which seeds embody, result from, and foster particular relations of power. If we ignore the complex compositions of seeds and focus only on their final function – increased yield or herbicide resistance, for instance – we miss a large part of the story. After all, 'the choices we make individually and collectively are heavily influenced by the way existing social institutions constrain the range of choices or set the implicit costs associated with the choices available' (Power 2000, 272). Seeds are more than simple tools or mere commodities.

Merging technology and commodity forms in neoliberalisation, as Monsanto suggests, provides goods from which to choose; however, the practices and implications of these goods are obscured, while problems are misapprehended. In other words, by assessing seeds as mere tools to consume as we choose, we ignore the refigurations of seed into commodities, savers into consumers, and seed saving into anachronistic practice. We also miss all the relations and questions that come with seed technopolitics. What did it take to get GE flax approved, and then deregistered? What might genetic engineering not enrolled with the corporate seed order offer? What would policies that valued saving look like? What possibilities are excluded in purchasing transnationals' 'best possible germplasm'? And so on.

Corporate control of seed ordering is a relational achievement – we all participate in reinforcing (or undermining) its consolidation. In Canada, since the 1970s state-facilitation of corporate strategies has been fundamental to amending regulations and enclosing, privatising and revaluing seed. Cultural change is part of this process – contract growing, marketability of research, public-private partnerships, farmers as manager-consumers, seeds as commodity-technologies, and industry entanglements in policy-making and implementation become accepted as part of how seed should be ordered. Moreover, Canada's movements toward neoliberalising are not isolated, but part of international trends. The corporatising seed order is an accomplishment of diverse actors at various scales, but it is an increasingly uneven one in which private capital takes primacy.

Transnationals have a particular vision of what today's seed order should be and have set about achieving it. Dupont/PioneerHi-Bred (2010: 1–3) explains that their vision 'requires quality inputs and aggressive farmer education, coupled with a regulatory environment that enables farmers to access innovative agricultural technologies.' Between this vision and Monsanto's endorsement of the 'best possible germplasm', several common arguments present themselves.

Ag-biotech argues that only by reordering through neoliberal processes will growers obtain 'the best possible germplasm' to deal with the challenges ahead, and only by leaving behind such outdated practices as seed saving will society succeed. Castree (2010a) summarises the arguments of neoliberal advocates with the acronym GEDDS, standing for growth, efficiency, development, democracy and sustainability. The advancement and manifestation of these arguments varies in practice, as Castree notes, but several have been demonstrated throughout this chapter as part of ongoing arguments of ag-biotech advocates for reordering seed-saver relations. Economic efficiency and individual choice, for instance, dominate appeals to growers to maximise their abilities to meet the challenges ahead by purchasing certified or GE seeds, although elements of environmental sustainability – in reduced herbicide applications or low-till production for instance – may also appear. Further, achieving food security and environmental sustainability depends, says ag-biotech and its supporters, upon technological innovation, which requires a facilitating regulatory environment. In this view, governments serve the public interest by promoting economic growth through fostering a flexible and business-friendly regulatory environment – through, for instance, reregulating seed registration, maintaining existing GE approval processes, and promoting GE LLP acceptance. The choice of corporations to offer one seed over another is not a neutral or unentangled decision.

What constitutes the best seed is not the same for all interested. Small-scale seed companies, public plant breeders, and seed varieties have all been lost in processes of building a corporate seed order (Podoll 2012). What is considered useful in the corporate profit sense is not necessarily the same as what is considered useful for local consumption or growing, for agrobiodiversity or for rural livelihoods (Shiva 1995). Lammerts et al. (1999) discuss the needs of organic growers in particular, explaining that growers should not continue to rely upon the commercial system for two primary reasons: first, varieties do not have required characteristics (such as high mineral efficiency, weed suppression, stress tolerance); and second, commercial breeding increasingly favours disallowed methods (like genetic engineering). In this context, as in earlier discussed certified seed debates, seed saving offers not only choices to growers, but likely better ones, combining cost savings and adapted, appropriate varieties. But this kind of practice – regrowing seeds rather than purchasing 'the best possible germplasm' – does not fit with the corporatising seed order.

Ag-biotech transnationals develop and maintain enabling technopolitical environments by mobilising their extraordinary capital to shape economies, discourses, policies and practices. These corporations own biological materials, intellectual property and technological infrastructure that supports (and depends upon) continuing commodification of seeds and expanding markets. The level of concentration of these resources allows transnationals remarkable influence. How transnationals pursue and fund breeding, which varieties they decide to market, how seeds are integrated into growing packages, where growers and breeders are contracted and under what conditions, are just some of ways in which corporations alter seed orders. But, of course, they cannot achieve or maintain this order alone – governments, growers (potential or actual savers), and seeds all participate.

Governments like Canada's, interested in pursuing commercialisation and competitiveness, facilitate corporatisation and neoliberalisation through forming collaborative partnerships with industry and its organisations. This does not mean that government does only, or all, that industry advocates. Moves to deregister GE flax, withdraw from certified seed inspections, or sign agreements preventing GURTs commercialisation, as examples, work against ag-biotech interests. However, overall, government and ag-biotech maintain cooperative relations, which often serve to increase industry influence while sidelining public concerns as well as accountability for governing.

The choice to consult rather than engage the public is understandable for resource-constrained agencies looking for useful input for efficient implementation, but governing should be more than this. The CFIA, for instance, acknowledges that '[s]eed policies are no longer only of interest to agricultural producers and the major national associations of the seed industry' but to 'a diverse and expanding community of stakeholders and interested parties' (CFIA 2006b: 17). The Agency also indicates that in feedback on seed policy revisions public concerns were expressed about globalisation, corporate control and profits, genetic engineering, and negative impacts on organic production. Moreover, resistance to certified seed revaluations, contestations of GE seed approvals, and mobilisations against Terminator seeds each illustrate publics insisting on their relevance. However, government's recognition of public concern has not translated into policy engagement. This would be less problematic if public discussions of the purposes and values that inform seed policies had occurred and were reflected in policy. However, that is not the case.

In addition to governments, growers participate in corporate seed ordering – as members of the public, seed consumers and, perhaps, savers. As seed consumers, growers, as Monsanto suggests, make choices to which we should pay attention; however, contra Monsanto, the choices and seeds involved are constrained. Consider, as an example, adoption of Monsanto's herbicide-resistant GE canola. In Canada, canola is by far the most grown GE crop (followed by corn and soybeans) accounting for about 5.4 million hectares in 2010 and approximately 80 per cent of all the canola grown in Canada (Brookes and Barfoot 2006; Lupescu 2011). Producers may have initially adopted these seeds, despite no yield increase, because chemical management is simplified, contract terms stipulate the variety, or cost savings are anticipated (Benbrook 1999, 2002; Clark 2003, 2005). Since these early adoptions, most growers interested in non-GE canola have given up trying to source uncontaminated seed due to limited variety choice and high levels of GE contamination, even in pedigreed seed (Friesen et al. 2003; Schmeiser 2005; Van Acker 2005). This combines with worries about inadvertent patent infringement implications, both before and after the Schmeiser case (see Chapter 4). Under these conditions, growers 'choose' to either switch crops or grow GE canola. Choices to adopt GE canola, themselves constrained, condition those of other growers – forcing market exit or GE canola production – and serve industry well by expanding commercial markets.

Aside from the consumption choices to which Monsanto implies it caters, and which it certainly constrains, growers make other kinds of seed choices. Choosing, for instance, to persist in saving their own seed or exchanging with neighbours, lobbying for economic risk evaluation of GE crops, and so on. While growers, and eaters for that matter, do make choices, these are entwined with existing assemblies involving all sorts of human and nonhuman actors – including seed.

If seed is the first link in food chains, then neoliberal seed is the first link in transnational seed ordering. The capacity of seeds to reproduce allows savers to work with them, developing diverse varieties and tendencies over time. But this capacity of seeds proves problematic for commodification – if seed can be bought once and then saved, it limits profit-making. In recent years seed corporations have developed more effective, though still imperfect, means to achieve seed commodification through biological and regulatory measures.

Enrolled in the evolving corporate seed order, seeds become isolated from their historic social and ecological communities, technologically altered, owned by virtue of IPR allocations, and revalued from living organisms to commodified containers for useful genetic traits. Borgmann (1984) suggests that this type of transformation be understood as a combination of technological and commodification processes, with resulting products – like seeds – becoming devices. He argues that as complex things (living or non) are remade into devices, they lose some of their 'thing-ness' or abilities to engage people in meaningful ethico-political relations. Terminator seeds come closest to Borgmann's device characterisation, but other (less extreme) forms of deepening commodification have been illustrated in this chapter.

These corporate seeds, adding to Borgmann's concern, embody a whole agrifood order dependent upon and reinforcing of expertise, chemical and commodified inputs, mechanistic production, and so on. Genetic engineering allows seeds to be further transformed to serve corporate purposes. A GE seed carrying, for instance, a bacteria's patented genetic sequence to express insecticidal traits in each of its cells is significantly different from in-field application of the same pesticide. Corporate profit relies upon seeds – without seeds new combinations of traits cannot be delivered and are not useful in the field. It is in this sense that I suggest, rather than a dualistic characterisation of seed as 'things' or 'devices', it is perhaps more useful to suggest that neoliberal, corporate seeds also embody and embed, facilitate and escape ordering, but in different ways – some of which limit seed saving.

Markets and corporations, governments and policies, savers and seeds become refigured through the corporatising seed order. So pervasive is this neoliberal ordering of seeds, particularly in the global North, that 'the existence of alternatives is hardly recognised, creating, overall, a potent and debilitating form of cognitive sclerosis' (Nazarea 2005: 11). Kneen (1993) refers to this phenomenon as 'distancing', a spatial and cognitive process through which people become disempowered and deskilled, resulting in inabilities to produce their own food or to eat well, or, I would add, to imagine or begin saving seeds. As distancing becomes normalised within seed networks, dependence upon and acceptance of corporate seed (and their technopolitical orderings) gain further purchase. We forget about the hows and whys, even though 'as technologies are being built and put to use, significant alterations in patterns of human activity and human institutions are already taking place. New worlds are being made' (Winner 1986: 11). Even in its most extreme forms and its most pervaded places, corporate seed reordering remains incomplete and disputed.

In this chapter disputes have been detailed relating to growers resisting certification schemes and persisting in seed saving, insistence on economic impact assessments for GE seed, and international rejection of Terminator seeds. Additional contestations have been noted regarding seed approval assessments, efforts to establish GE-free zones, demands for redressing GE contamination, and rejections of GE seeds by consumers and states. Further, disagreements exist among transnationals. The acquisitions, mergers and licensing agreements among top corporations unquestionably consolidate corporate control. But in the scramble for market share, some transnationals question others' practices; and, even if an underlying logic of neoliberalisation and corporatisation is shared, such discord raises questions about corporate consolidation and price controls.

Further, there are regional and crop differences in market penetration and concentration that leave gaps in corporate seed ordering. GE crop adoption and acceptance varies – for instance, the European Union's approach, research and consumer responses diverge from those of North America (Jasanoff 2005; Murphy and Levidow 2006). In addition, not all crops are equally controlled or researched by corporations – soybean and canola, for instance, receive more attention than flax or wheat (though this is changing). While limited research and development means fewer new varieties, it also avoids monopolisation that often comes with corporate attention, and GE contamination. This gap also keeps open opportunities for breeding and continued saving – restricted opportunities certainly, but still options.

Moreover, the commercial seed market is only one way in which growers obtain seed. Seed saving continues to be an important practice for much of the world, with estimates that 70–80 per cent of seed is still saved worldwide (ETC Group 2008; Monsanto 2009: 49). And, as illustrated, even in the global North – where ag-biotech has its strongest hold – commercial growers find the practice useful and persist. Finally, seeds themselves contest the corporate seed order. Even as seed bodies and valuations alter to serve private capital, seeds escape and continue – not all plant cells transform, not all GE traits express as predicted, GE seeds and pollen contaminate, and seeds endure in fields as well as through saving practices.

The strength and resilience of the evolving corporate seed order and its impacts on seeds and savers, as well as wider agrifood, should not be ignored. Neoliberalisation with seed corporations increasingly dominates contemporary seed orderings, especially in the global North. In this chapter I have shown several strategies of industry and collaborating governments to maintain and expand a corporate seed order. It is clear that these efforts significantly alter seed orders, but this process is neither complete nor uncontested.

It is not, I maintain, our task to abandon as essentially 'good' or 'bad' particular methods, technologies or strategies, but rather to interrogate their relations and implications, deciding together which to follow and support, which to abandon and exclude, which to revise and resubmit. This is a theme that continues in the following chapter, which examines intellectual property rights and, as counterpoint, farmers' rights.

Chapter 4
Configuring Rights

Seedy Saturdays are social occasions filled with conversation – people catching up, planning for the coming year, reflecting on what they have learned from last year. They swap and buy seeds, check out resources and informational booths, go to workshops, grab a bite to eat. I enjoy it immensely. It's spring, or almost spring, and people are inspiring each other, their shared pursuits, and the possibilities of the coming year – new seeds and techniques to try, old ones to keep improving.

Fig. 4.1   Seedy Saturday in Victoria, British Columbia

But in 2005 things felt different. There was tension. It wasn't that the fascination, anticipation or celebration disappeared. Far from it. But there were cracks, and not from a tough season or similar problems, which can happen at these regional events. It was more widespread.

I overheard people talking about how the government was going to make seed saving illegal, or at least harder. I heard it expressed at the booths, in the aisles, even while snacking in the food areas; 'those proposals aren't right', 'they're crazy', and similar phrases were uttered over and over. In late 2004 changes to plant breeders' rights were proposed by the Canadian government, and savers were reacting. There was a petition going around to stop the changes, and to encode seed saving clearly in legislation. There were pamphlets circulating with information on the changes. I even saw a couple of declarative t-shirts defending seed saving: 'Save seed, not greed!' There was resistance in the air, even rebellion – 'we'll save seeds anyway!' Not everyone knew about the policy proposals of course, and the details were often shadowy. Not everyone cared either – they were not affected. But word was spreading and some people were seriously agitated.

Intellectual property rights (IPR) such as plant breeders' rights (PBR) – and the governing practices of which they are a part – matter in how seeds, seed savers and their relational others live with and in their worlds. 'What is at stake', remind Fowler and Mooney in their discussion of plant-related IPR, 'is the integrity, future and control of the first link in the food chain. How these issues are decided will determine to whom we pray for our daily bread' (1990: xiv). IPR attempt, through legislation, to order the lively collaborations of seeds and savers. Through encoding, supporting, promoting and implementing seed-related IPR, domestic and international policies enable particular socionatural arrangements while redirecting challenges to these forms. More specifically, seed-related IPR facilitate the emerging corporate seed order detailed in the previous chapter by creating the conditions for privatisation. But IPR are not simple. They are part of ongoing and complicated negotiations and governing, and while they are abstract legal constructions, they make significant practical and ethico-political differences.

In this chapter I consider the connections between IPR and seed saving, using a particular illustration of IPR governance. I look closely at plant-related IPR as they manifest in Canada in a particular moment, making connections to related international treaties and relations. As counterpoint to these protections, I consider farmers' rights – rights to save, exchange, sell seed – again, as they present within Canada in relation with international agreements.

Intellectual Property Rights Taking Shapes

Intellectual property rights is a generic term used to refer to a group of legal protections – patents, trademarks, plant breeders' rights and copyrights – intended to protect individual creators/inventors from unauthorised copying or use of their work. Walking down the supermarket aisle provides plenty of examples: trademarks on brands and formulas like Coca-Cola; geographical indicators for regionally differentiated products like cheese and wine; patented shapes and methods for items like mustard squeeze bottles or processes used to make frozen pizzas. The same is true of browsing in plant nurseries or reading farming product guides. IPR pervade agrifood worlds.

In relation to saving seeds, PBR and patents are the IPR of most interest. Disputes over seeds and plants are not new; there are plenty of examples throughout history, but the battles over ownership and exchange of seeds (and other plant materials) as plant genetic resources (PGR) have intensified dramatically in recent years. IPR form part of the arsenal employed in these battles, defining and assisting legal privatisation of seed and control of seed-people relations. While the specifics of IPR vary from state to state, there is surprising consistency in their basic forms. In general, both PBR and patents provide protections to persons – legally including corporations – that prevent others from using protected material in certain ways for fixed periods. But there are differences. Two of the most significant relate to the designated object of protection and the allowed exemptions; PBR apply to varieties, whereas patents offer wider claim choices, and exemptions are generally more restricted with patents than PBR.

IPR agreements are part of continuing disputes and shifting alliances among the various states, social movements and corporations trying to influence how seeds, and their relationships with humans and other nonhumans, are governed. Differences in IPR and relative positions of advocates in negotiating forums, are reflected in discontinuities and continuities in approaches to seeds and growers among agreements. Indeed, the legislative and regulatory frameworks of other countries can be more prohibitive or more permissible of seed saving in comparison to Canada (*cf.* France, India, United States) but this does not lessen the challenge Canadian seed savers (and their supporters) face in defending and building seed saving networks, whether through farmers' rights or something else. Though IPR may seem static or unquestionable in specific policies and agreements, this is not the case.

IPR and the agreements of which they are a part are themselves assemblages – never complete or finished in the ordering work they do. Intellectual property rights, then, take particular shapes as they work at shaping seed orders.

Intellectual Property Rights in Canada and Beyond

In Canada intellectual property rights for seeds were not quickly adopted. Industry demands for IPR on plants began in the 1920s, but at that time agricultural research had a strong history located within the public sector, with a clear understanding that the development of better seeds was beneficial for all Canadians. This tradition combined with ethico-political concerns – about owning lifeforms and contributing to increased corporate control of agrifood – to hold off plant-related IPR. In the mid-1970s, however, things began to shift. At this time, SeCan was created by a consortium of seed companies, seed growers and government agencies to administer exclusive licences for the multiplication, distribution and marketing of seed varieties considered useful for Canadian agriculture (Slinkard et al. 1995). By charging membership levies and administering royalties, SeCan began moving breeding costs to growers and transforming breeding into a business with grower-clients, rather than a public good based in public breeding (Kuyek 2004a).

In the 1980s this process continued with federal governmental priorities reflecting a growing belief that the private sector, including biotechnology corporations, was fundamental to developing Canada's agrifood sector. During this time several incentives and disincentives coalesced to ensure development of the private seed sector including tax breaks for biotechnology research, and tying public research agendas to matching funds and marketability (Industry Canada 1993, 2007; Moore 2002). The move to public–private partnerships in breeding, formalised by the Matching Investment Initiative in 1995, makes it difficult for public scientists to research outside of industry-identified priorities because industry not only has close relationships with government, and influences breeding agendas, but it also controls funding potential. This arrangement means, for instance, that research on sustainable agriculture and seeds suited to these systems is often ignored in favour of work on more broadly marketable varieties (COG 2005).

By 1990 the federal government was convinced that IPR for seeds were instrumental to ensuring the competitiveness and innovation of the agrifood sector, and adopted plant breeders' rights.

Plant Breeders' Rights and UPOV

In Canada the *Plant Breeders' Rights Act* (1990) reflects one of the two major international agreements dealing with seed-related IPR: the *Union Internationale pour la Protection des Obtentions Végétales* (UPOV) agreement. UPOV, initially signed in 1961, was the first international agreement connecting IPR with plants. At the urging of commercial plant breeders, UPOV was formulated by a few European states to allow breeders some measure of control over the commercial seeds they bred, and to harmonise various national approaches in order to promote the commercial seed sector (Dutfield 2011). Despite early discussion of plant-related patents as an option, UPOV61 representatives believed plants – as living beings with reproductive capacity – required a different approach than technological inventions and instead referred to 'plant breeders' rights' that provided less encompassing property protection than patents. Since 1961 UPOV has gone through several revisions with reworked agreements in 1972, 1978 and 1991. The last two versions are the most commonly used by states, and are the ones charted in comparison to TRIPs-compliant patents in Table 4.1. Canada's PBR parallel UPOV78.

Canada's PBR Act and UPOV78 stipulate that PBR apply to a plant variety, or a group of plants that is distinguishable from others in the same species – different kinds of lettuce or apples for instance. To be awarded rights, breeders must submit documentation showing the variety to be new, distinct, stable and uniform. What this amounts to is meeting parameters regarding a variety's commercial history (newness), difference from other known varieties (distinctiveness), ability to repeatedly reproduce similar forms (stability), and comparative genetic uniformity. With these criteria, landraces and most farmer-bred varieties are excluded.

Once PBR are granted in Canada, the assigned rights holder has legal control over the multiplication and sale of the variety for a period of up to 18 years, and, therefore, may license and charge royalties for particular uses of the variety's seed. The PBR Act prohibits seeds protected under the Act from being propagated for commercial purposes without permission from the rights assignee. However, a breeders' exemption allows for research activities to continue with protected varieties, under specified conditions. Non-commercial reproduction remains undetailed, and farmers' rights or seed saving are not mentioned, leaving these options open. Most countries with PBR (also called plant variety protections) allow some form of private and non-commercial saving activities – with interpretations varying widely from permitting sales of saved seed to only allowing on-farm reuse. Many states – including Canada – continue to use UPOV78 as their framework, but a transition is underway and new UPOV members are required to accept UPOV91.

In Canada, although PBRs are administered by the Canadian Food Inspection Agency (CFIA), enforcement is left to rights holders – increasingly private industry. In the first ten years of the PBR Act industry recorded 40–50 violations (Heads Up 2000). In order to enforce the Act more efficiently, seed industry interests established the Canadian Plant Technology Agency in 1999. Private contractors, sometimes retired police officers, are hired to pursue cases of potential infringement, a strategy that some farmers consider purposefully intimidating and misleading when investigators identify themselves as ex-police (Kaskey 2012; Schmeiser 2005). In October of 2006 the Agency stated that so far that year it had stopped sales of protected seed valued at approximately CAD2 million dollars (Investment 2006). While this valuation may be overstated as a means of legitimating the agency's role and effectiveness, a more conservative estimate would still represent a scaling up of industry's enforcement of PBR.

In 2004, as part of collaborative efforts with industry to regulate seed in more flexible and commerce-friendly ways, the CFIA proposed changes to the PBR Act to bring national legislation into agreement with UPOV91. The move from UPOV78 to UPOV91 would not be insignificant. In general, UPOV91 extends breeders' rights and further constrains exceptions. In Canada's proposed changes to the PBR Act protections are newly offered to harvested material and essentially-derived varieties, include a wider range of activities (reproduction, conditioning, stocking), and apply for longer durations. Further, the prohibition on combining patents and PBR – called double protection – is eliminated. The exemption for research remains, though it is more restricted. Crucially for those interested in seed saving, including all those people at Seedy Saturdays, the ability of growers to save protected seed alters with UPOV91.

Under UPOV91 a growers' exception (called 'farmers' privilege') has been made optional. If Canada includes the clause, farmers would be allowed to save and reuse a protected variety on their own farm for their own purposes (CFIA 2006b). If, however, the clause is not adopted, growers could not legally save or reuse protected seed for any purpose unless authorised. During the consultation phase for the proposed changes, it appeared that the CFIA advocated not including the clause; however, after public response it clarified its position as supporting the exception. Even with the inclusion of a farmers' privilege clause, however, UPOV91 constrains seed saving by legally encoding it in emaciated form – only saving for personal use on-farm would be allowed and only as a privilege, rather than a right or norm.

Based on all of these policy changes, adopting UPOV91 strengthens IPR, bringing PBR much closer to the levels of control afforded by patents, though still with some limited exceptions. For countries that do not offer patents on plant varieties but are members of UPOV, this is particularly significant. Moves of this kind, which include progressive prohibitions on seed saving, facilitate privatisation processes and corporate control of seeds.

In the face of public outcry despite limited publicity of the changes, the CFIA put aside proposed changes to the PBR Act. However, government and industry representatives continue to argue that Canada misses out on opportunities by not proceeding (CFIA 2002; CSTA 2004b). Further, attempts have been made to use methods other than legislation to achieve UPOV91 conditions, including recent negotiations of a trade agreement with Europe (NFU 2012a). Moreover, industry argues that even UPOV91 measures are insufficient and that additional strategies are needed. As Bill Leask (1998: 95), Executive Vice-President of the CSTA at the time, explained:

> Canada's Plant Breeders' Rights still have two significant exemptions which are very problematic to the industry – a research exemption and a farmers' exemption. The research exemption means other researchers can use your material to develop

new products. The farmers' exemption allows farmers to plant their own seed on their own holdings. Many of the new players in our industry find these exemptions too onerous. So, Canadian breeders have moved to other forms of intellectual property protection, such as patents. Now contracts are used to plug the farmers' exemption – the ability to farm and plant back his [sic] own seed.

Contracts reinforce IPR (as noted in Chapter 3) or 'plug' PBR exemptions for already released varieties (or those that do not meet PBR criteria), but, as this quote indicates, the IPR preferred by industry are now patents.

Patents and TRIPs

Plant breeders' rights cater specifically to the needs of breeders, but patents offer another, quite different, approach to plant-related IPR in Canada and internationally. In addition to belonging to UPOV, Canada is a member of the World Trade Organization (WTO) and, therefore, committed to the Agreement on the Trade-Related Aspects of Intellectual Property Rights (TRIPs). Unlike the continuity between Canada's PBR Act and UPOV78, Canada's seed-related patenting meets but does not parallel TRIPs provisions. The following section, therefore, deals first with the TRIPs framework and then Canada's particular approach to seed-related patenting.

TRIPs
TRIPs serves as a connective element between two international regimes – IPR and trade – and was incorporated into the multilateral agreement that established the WTO in 1994 as part of a product bundle for joining states. Negotiations over its particular measures and form have been ongoing since 1986 and, in general, have been led by representatives of developed countries. The US takes a particularly strong stance in favour of IPR – a position increasingly mirrored by the Canadian government.

TRIPs establishes minimum standards for member countries to offer and enforce IPR. In the one paragraph relating to plant materials, it specifies that member states may 'exclude from patentability':

> plants and animals other than microorganisms; and essentially biological processes for the production of plants or animals other than non-biological and microbiological processes. However, Members shall provide for the protection for plant varieties either by patents or by an effective *sui generis* system or by any combination thereof.

This text distinguishes between plants, their essential biological processes, and plant varieties, differentiating protections required. In TRIPs, although some organisms and processes may be exceptions, members are required to adopt legislation that assigns intellectual property rights to plant varieties in the form of patents and/or an effective *sui generis* (or 'of its own kind') system. Without appropriate IPR protections, member states are deemed to be restraining trade, and any restraints on trade by member states, as outlined in other sections of WTO agreements, must be justified. Any disputes among members over such issues may be addressed through the WTO's binding dispute settlement system.[1]

In contrast to other IPR outlined in TRIPs, a standard is not set by reference to another international agreement. Existing plant protection agreements such as UPOV78 or 91 might be considered *sui generis* systems, but since TRIPs does not cite these as standards, states are left to formulate the specifics. However, if states allow plant-related patents, TRIPs offers clarification that must be accommodated.

According to TRIPs, patents may be applied when 'inventions' – products or processes – are novel, non-obvious and useful. Because the word 'invention' remains undefined, states variously apply patents to: plant varieties; whole organisms, like seeds or plants; plant parts, such as gene sequences, genes, and so on; and/or to breeding methods. Further, negations of novelty and obviousness range widely from requiring written evidence to any evidence of prior public knowledge. These interpretations are particularly relevant for considerations of traditional knowledges and practices, which have been largely excluded from IPR allocations and exploited for the benefit of others (Dutfield 2009; Mgbeoji 2006; Shiva 1997).

In comparison with PBR, patents come with extensive exclusive rights and limited exceptions (see Table 4.1).[2] Most state laws allow for some level of research exemption in relation to patents, but they are more restrictive than those of PBR. For example, in the US an unauthorised breeder is prohibited from crossing patented seeds to produce other improved varieties (Correa 2000), but PBR permit this kind of activity. In addition to limited accommodations of research, patents rarely permit growers to save seed, even for their own purposes. Patent protections are reinforced for growers through production contracts – called technology use agreements by corporations – that make use prohibitions clear, and subject not only to IPR protection but also to contract law. With or without contract reinforcement, the combination of an extensive scope of activities and limited exemptions allows patent holders more control over licensing others than do PBR. Overall, the limited attention and detail TRIPs gives to seed-related IPR belies the significant and controversial role it and proprietary rights play in agrifood.

---

1   Disputes relating to agrifood IPR citing TRIPs have so far included trademarks, geographical indicators, and patents on food stuffs, pharmaceuticals and agricultural chemicals.

2   Compulsory licensing is used by some countries to allow by-passing of licensing requirements of patented inventions, often with public good justifications. Some food and medicine products, for instance, have received this treatment in Canada.

Table 4.1 Comparison of IPR in UPOV78, UPOV91 and TRIPs

	PVP under UPOV78	PVP under UPOV91	Patents under TRIPs
Object	Plant varieties	Plant varieties	Inventions
Criteria	New to market Distinct Uniform Stable	New to market Distinct Uniform Stable	Novel Non-obvious Useful
Duration	Minimum 15 years 18 years – vines/ trees	Minimum 20 years 25 years – vines/ trees	Minimum 20 years
Protected activities	Producing for commerce Offering for sale Market	Re/Producing Conditioning Offering for sale Selling/Marketing Exporting/ Importing Stocking	For Product: Making Using Offering for sale Selling Importing For Process: Using Any listed for products if obtained through patented process
Protected material	Propagation material	Propagation material Essentially derived varieties (partial) Harvested material (partial)	Product Process
Breeders' exemption	Yes	Yes	Up to state Not often provided
Farmers' exemption	Implicit	Optional, restricted	Up to state Not often provided

Adapted from: Dutfield (2008); Helfer (2004)

Patents in Canada

With the TRIPs provisions in mind, what does the Canadian approach to plant-related patents look like? On the one hand, Canada does not permit patents on 'higher life forms', which include animals as well as seeds and plants (see *Pioneer Hi-Bred Ltd. v. Canada 1989*; *Monsanto Canada Inc. v. Schmeiser 2004*;

CBAC 2002), but on the other, proteins, genes and processes are patentable under Canada's Patent Act assuming they meet the criteria (see Patent Appeal Board of Canada 1982). These distinctions in Canada differ from the approach in the US, for instance, where patents are allowed on varieties, organisms, parts of organisms and related processes. To complicate matters, while the *Patent Act* (1985) is administered by the Patent Office within Industry Canada, the seeds within which patented parts are incorporated are still legislated by the *Seeds Act* (1985) and its *Regulations* (2012) under the CFIA.

The legal division between higher life forms and lower life forms (and non-organisms) may seem intuitively clear, but practically it proves tricky. Disallowing one, while allowing the other, means that the distinctions can become meaningless. This possibility became apparent with the controversy surrounding a lawsuit ultimately heard by the Supreme Court of Canada.

In 2004 the Supreme Court of Canada changed the landscape of IPR and GE contamination in Canada with their now famous ruling on *Monsanto Canada Inc. v. Schmeiser*. In 1998 Monsanto Canada Inc. accused Percy Schmeiser, a rapeseed/canola farmer from Saskatchewan, of infringing on their patent for a gene that confers glyphosate-resistance. Monsanto's inspectors had tested canola in Schmeiser's fields and found the presence of Monsanto's patented gene. The GE canola had been introduced into the region in 1996, but Schmeiser claimed that he had not bought or deliberately sown it. Instead, Schmeiser said he relied on his own saved seed, which he had adapted and selected over the previous 50 years of farming canola, and that any presence of the GE canola in his fields must be due to contamination (Schmeiser 2005).

After the case worked its way through the courts, the Supreme Court of Canada ruled against Schmeiser in a 5–4 decision. The ruling states that Schmeiser infringed on Monsanto's patent by 'cultivating a plant containing the patented gene and composed of the patented cells without license', and that, regardless of whether he commercially benefited from the gene's presence, he 'used' the patent within a commercial context by selling the crop. In other words, Schmeiser was guilty of patent infringement because he cultivated, saved and sold canola that contained Monsanto's patented gene without Monsanto's consent; whether he knowingly did so was irrelevant. Schmeiser lost his crop – contaminated and non-contaminated canola – but Monsanto Canada Inc. did not receive any remedial payment since Schmeiser did not benefit financially from the presence of the gene (he sold his canola for the same amount that he would have received for non-GE canola).

While seed-plant organisms may not be patentable *de jure* in Canada, it seems they have become so *de facto*. While the SCC decision pleased the seed industry (CSTA 2004b), others were more critical; for example, Pat Mooney of the ETC Group stated in the wake of the ruling that 'Monsanto has won an inflatable patent today. They can now say that their rights extend to anything its genes get into, whether plant, animal, or human. ... Under this ruling spreading GM pollution appears to be recognised as a viable corporate ownership strategy' (ETC Group 2004). Although the ruling made it clear that the patent was on the gene and not on the whole plant, the result of the ruling is that the organism is protected as Monsanto's property by virtue of the fact that the gene exists in the plant.

In addition, the court's interpretation that intent is irrelevant means that farmers become more vulnerable to patent holders. When threatened with legal action, most farmers choose to settle out of court, paying patent holders – usually corporations – rather than face the possible loss of their farm and/or crops (CFS 2004, 2007). It is unlikely that patent holders will pursue all farmers who have GE-contaminated fields as violators of IPR; however, it is clear from the Schmeiser case and increased PBR enforcement that corporations will more frequently be pursuing (or threatening to pursue) alleged infringements – intentional or otherwise.

The Schmeiser case points to more than a shift in IPR legislation, connecting as it does with GE contamination and seed saving. In an effort to make positive use of the verdict, some growers argue that if patent holders have rights to plants by virtue of their patent being contained within them, then patent holders have responsibilities as well – particularly for ensuring contamination does not occur, for cleanup, and for compensation. For instance, in 2002 several organic farmers from Saskatchewan pursued a class action lawsuit against Monsanto and Aventis on the basis of the contamination of fields by GE-canola.[3] In 2008, when the class action procedure was denied by the courts, one of the involved growers made it clear that their fight was not ending, commenting that growers will continue to 'challenge Monsanto and Bayer for the liberty, freedom and right to grow GMO free crops. We want to be able to save and use our own seed' (as in OAPF 2008). More recently, several Canadian agricultural organisations and growers have joined the Organic Seed Growers and Trade Association in the US in their ongoing suit challenging Monsanto's patent protections on GE seed. In this way, efforts to clarify liability for GE contamination continue not only in public lobbying for regulatory change (see Chapter 3) but in legal forums as well.

Connecting forms and forums

The differences and complications of Canadian plant-related IPR are a reflection of domestic and international relations. Seed and IPR in Canada come together through federal government collaborations with industry, which combine with international interactions – particularly those involving influential

---

3   For legal case reviews see Phillipson (2005); de Beer (2007).

trading partners. Canada faces the US, a powerful IPR advocate, as its major trading partner. Canada has not yet harmonised with the IPR strategy of the US by, for example, enacting legislation allowing patenting of whole plants and/or animals. However, Canadian legislation and international advocacy increasingly echo US approaches. As exemplar, Canada strongly argues, along with the US, against the inclusion of farmers' rights provisions in domestic and international policies despite public pressure domestically to retain seed saving (Richards 2007; Wells 2005; Whatmore 2002).

IPR are still primarily state-based as they are adopted, allocated and largely enforced at national levels, but this is changing. As discussed, TRIPs and UPOV present members with options for ordering seed–people relations through specified IPR systems. TRIPs requires minimum standards of plant variety protection of its members, providing optional enforcement through the WTO's dispute mechanism. In another forum – the World Intellectual Property Organization, a special agency of the United Nations – worldwide harmonisation of IPR standards is the goal. IPR applications, for instance, can now be made simultaneously in several countries through WIPO arrangements. In addition to IPR being extended and standardised through international treaties and organisations, bilateral and multilateral agreements are increasingly used to push beyond TRIPs specifications – earning themselves the label 'TRIPs-plus.' Adding to their more secure and broadened IPR, these agreements often incorporate their own dispute resolution mechanisms furthering shifts from state-based administration and public scrutiny. North American and European governments have both been criticised for their use of direct pressure tactics and aid packages in furthering IPR and harmonisation (GRAIN 2005a, b).

Moreover, the multiple arenas for IPR negotiations allow those states with financial and legal capability to 'forum shop' – choosing the organisation or agreement that best serves their interests at that point. This can be seen, for instance, when difficulties with pushing for strong IPR protections within WIPO led IPR advocates to shift their agenda to trade negotiations – resulting in the formation of TRIPs in the WTO – a move, in its turn, followed by a recent returning to some WIPO treaties and more direct developments of bilateral and multilateral agreements because of the perceived limits of WTO effectiveness (Rajotte 2008). The forms and forums of IPR continue changing, advancing proprietary options.

Intellectual Property Rights Making Differences

What differences do IPR make? Advocates argue that the incentives and rewards provided through IPR foster innovation, which benefits national economies and society at large. For instance, the Canadian government, informed by the seed industry, initiated the PBR Act as an effort to: stimulate private investment in breeding programs; improve the ability to obtain foreign varieties; protect Canadian commercial interests abroad; and improve plant varieties for public benefit (*Plant Breeders' Rights Act* (1990)). The government-commissioned, ten-year review of the PBR Act suggests that PBR have, overall, been beneficial for Canadian agriculture by increasing access to foreign varieties and private investment in plant breeding (CFIA 2002). Further, this report recommends Canada bring the PBR Act into accord with UPOV91. However, the success of IPR at advancing plant breeding and the common good in Canada is disputed, with critics citing breeding limitations, increased seed costs, reduced seed choice, increased enclosure of seed, and greater corporate control through privatisation (Kuyek 2004a; Phillips 2008; Prudham 2007). Justifying IPR is more difficult than advocates suggest.

Designations in IPR are not neutral; rather, they reconfigure power and human-nonhuman relations. Much time has been spent examining the differential social and environment effects of plant-related IPR, often in relation to food and agriculture and to conflicting global North and South interests (see Andersen 2008; Blakeney 2009; de Beer 2005; Jasanoff 2005; Louwaars et al. 2005; Shiva 2000; Tansey and Rajotte 2008; UPOV 2005; Wong and Dutfield 2011). In this context, the complicated connections of seeds and IPR bring with them issues of nature-culture dualisms, epistemological differences, and social justice concerns about biopiracy or prospecting, farmers' livelihoods, differential state capacities and political influence, and corporate manoeuvring. With a focus on how IPR more directly connect with seed and saving in Canada, I consider how they make differences in two inter-related ways: by establishing and judging governable objects; and by authorising owners and activities.

Establishing and Judging Objects

The first difference made as seed saving and IPR meet is about transforming mutable organisms – like seeds – into stable artefacts to be governed. This involves two moves for seeds: first, they – and various parts of them – are redefined to become governable objects; and second, these isolated objects are remade to be static with the application of particular criteria to ease governability. These moves together reconstitute seed-plants in ways favouring private ownership, and re-establish nature-culture dualism.

First, IPR establish the object to be governed – cutting into the materiality of seeds (see Whatmore 2002). This kind of cut presents in both plant breeders' rights and patents, though their particularities differ. Plant breeders' rights designate plant varieties – or the plants of species' subgroups – as their object. In contrast, patents govern an array of objects – varieties; whole organisms – plants or seeds; parts of organisms – genes or genetic sequences; and the processes used in reconstituting these objects into inventions. Patents, therefore, make deeper cuts into the flesh of seeds, but both types of IPRs work to categorise and divide.

Second, those objects newly made by IPR must meet particular criteria to become governable. For plant breeders' rights to apply the variety must be new, distinct, uniform and stable, while with patents the object must be novel, non-obvious and useful. As previously noted, in effect this means objects must have no previous commercial history in the legal jurisdiction, and applicants must have altered the object in a useful and reproducible way. While each criterion must be met, seed IPR hinge on individuals transforming natures. As Strathern (2001: 9) explains in relation to patents, 'the general rationale is that patents cannot apply to any interpretation of manipulation of natural processes that does not require the specific input of human know-how resulting in things which did not exist before.' For both sets of IPR, only those objects sufficiently transformed count, the rest remain unimproved natural resources. Further, each requires scientific documentation and assessments. In this way, plants-seeds are considered raw materials of nature until they are developed through scientific, documented processes, whereupon they – as varieties, organisms, or their traits and parts – become products of intellectual labours and therefore governable with IPR. I do not suggest here that labour is not involved in breeding new varieties, but that in this process of establishing the objects of IPR, seeds' capacities and histories are erased, with new valuations changing bodies and relations through legal encoding.

Humans are centred as the source of change here, since it is particular human knowing that transforms seeds into useful objects. It is in this sense that Shiva (1995: 275) argues that IPR – particularly patents – do violence to seeds. By denying seeds their self-organising and self-reproducing capacities, by remaking them into static, dead artefacts transformed only through human scientific intellect, IPR offend seeds. Further, IPR ignore the ongoing sociality of seeds as participants in complex associations with people and other nonhumans. IPR criteria, then, judge objects, transforming those that pass into durable legal entities, reconstituting boundaries between natural resources and cultural inventions such that only 'improved' objects are eligible for protection while others remain in a separate, ungovernable, realm of nature.

The increasing reliance upon and requirements to follow international standards for IPR suggests this remaking of seeds is a universal, inevitable process; however, object definitions and criteria are complicated and their meanings are diverse, contested and changing. IPR are continually reinterpreted and reforming as legal orders try to adapt to changing relations. As de Laet (2000) notes, IPR are themselves changeable objects as well as agents of change. Canada's patents are not the same as those of the US, despite harmonisation efforts, and they have altered over time. Further, there are tensions in Canada's shifts toward stronger protections. These are evidenced by contestations regarding harmonisation with trading partners and corporate control – for instance through court challenges, as well as advocacy for, and continued practices of, seed saving. Finally, as discussed, the rights, prohibitions, and implications of IPR alter as their forms change – for example, if the proposed changes to the PBR Act were approved.

Re-established boundaries between nature and culture have also been unsettled. IPR have been awarded, for instance, on unaltered objects. In some cases this has brought forth new kinds of IPR – justifying isolation as sufficient alteration in the case of some states' patents for instance, and therefore further blurring legal distinctions between discovery and invention. In other instances, indigenous and farmers' varieties, or their parts, have been claimed and individuals – persons or corporations – awarded IPR. Legal disputes of these IPR have sometimes succeeded. Patents have been revoked for neem's fungicidal qualities, turmeric's medicinal qualities, the 'enola bean' and others (Bullard 2005; CGIAR 2008; Raghavan 2000). But legal challenges are difficult, lengthy, and expensive, and therefore many cases of this kind are unlikely. Interestingly, successful challenges have relied on providing evidence of prior use and knowledge, disputing individual claims of invention. In these cases, objects are not simply returned to 'nature'. Their enculturing is first explained otherwise, and only then are they ushered back to rejoin the unprotected – cultured, but not by that individual, and not in the right way.

Finally, the cuts made by IPR into seed bodies are messed with by the seeds themselves. Seeds continue to reproduce, associate and move beyond expected bounds, despite IPR's orders. Seed mobilities and mutabilities of this kind are perhaps made clearest as IPR-protected, genetically engineered crops travel on their own to become contamination or super-weeds. But in other instances, seeds enrol humans – as travel agents and as collaborators in seed saving, indifferent to the expectations and rules of IPRs.

Authorising owners and practices

The second difference made with seed-related IPR involves authority formation and assignment of rights, and the socio-economic implications of these designations. Both patents and PBR designate individuals – persons, institutes or corporations – responsible for the step just previous to application as appropriate rights' holders. It is then, only the last phase of development and only the controller of that phase that matters. In this formulation, the history of knowing is erased, along with the diverse contributions of multiple subjects and objects. Long traditions of working with seeds and participation of the seeds themselves are lost in favour of individual achievement rewarded with property ownership. With IPR, it is the inventing individual that deserves to decide future use. This assignment and authorisation comes with significant affects at various scales. I focus primarily in this section on the impacts for breeders and savers, particularly within a Canadian context. But seed-related IPR – and Canada's relations with them – reach further than this.

IPR and their affects are layered within the existing seed order and interpreted differently from differing positions in that system. Building on genebanking, commercial seed practices and state sovereignty principles, corporations and governments use IPR to enclose seeds and saving, furthering privatisation to increase profits and market share. This is done without necessarily acknowledging the origins or histories of materials – most of which come from farmers' fields and wildlands of the global South (Borowiak 2004; Mgbeoji 2006; Robinson 2010; Shiva 1997). The expansion and strengthening of intellectual property rights, and the influence of a small number of industry and legal experts in writing the rules, reinforce and refuel feelings in global South countries of exploitation by the diversity-poor but technology-rich global North (Dutfield 2009; Shiva 1997; Tansey 2002; Whatmore 2002: ch.5). In addition, in international and bilateral relations particular IPR legislation – UPOV, TRIPs, TRIPs-plus – are promoted as signals that adopting countries support new varieties and economic development; however, these arrangements are often questioned on grounds of transparency, equal participation and appropriateness (Dutfield 2009, 2011; Tansey and Rajotte 2008).

Not surprisingly, global North countries, collaborating with ag-biotech officials and building on existing socio-economic inequities, reconstitute seed orders to their advantage. It is in Canada's interest to ensure state sovereignty over the seed within its genebanks, and close connections between the state and industry explain much of the recent advocacy for strong IPR. Although global IPR issues such as biopiracy and global South food sovereignty were of concern to some of the savers who participated in this research, primary focus fell on IPR implications for public breeding programs and, most strongly, practices of saving seed in Canada – subjects to which I now turn.

Breeding with Intellectual Property Rights
Plant breeding both shapes and is shaped by IPR. As noted, it was plant breeders and their companies that pushed plant breeders' rights forward with the initial signing of UPOV. Breeding work includes doing basic research through to production (sometimes including marketing), and involves a wide-range of techniques as part of customary selection and reproduction, hybridisation or biotechnology. IPR, in relation to plant breeding, are meant to recognise innovation, provide return on investment and contribute to the public good through advancing knowledge and useful products. But IPR benefit some breeding programs and some breeders more than others (Jasanoff 2005), with patents being particularly problematic for work by public institutions and even smaller companies.

There are two associated conflicts with patents and breeding: first, access to material and information; and second, diversity of material and varieties. One of the worst implications of patents from the perspective of a plant breeder is the lack of a breeders' exemption (Myers 2012). Plant breeders must now deconstruct any proposed research into possible component organisms and methods in order to assess and deal with any IPR. Since research can involve genetic material and research methods protected under many different licensing agreements, establishing 'freedom to operate' can be quite difficult and costly (Kloppenburg 2004; Wright and Pardey 2006). Moreover, exclusive controls are extended through 'strategic patenting' practices, which involve denying licensing and prolonging patent rights through reapplications of tweaked products. This strategy is increasingly practiced by transnationals that hold the vast majority of patents, and weak application of criteria by governments makes this process even easier (Louwaars et al. 2009). The complexity of IPR assignments exacerbates this situation, making it difficult for breeders to even know which licenses are required to proceed. Indeed, in some cases it is easier for corporations to buy a company to gain its patents – an option not available to public breeders – rather than attempt to gain or pay for licensing,. Even with PBR, which provides some exemption for breeding and research, some breeders display reluctance in sharing what could turn into lucrative research (Kuyek 2004b).

Breeding with IPR becomes an expensive and fraught endeavour, sometimes prohibitively so for public breeders. In recent years public plant breeding budgets have decreased, and the move to public-private partnerships has not ameliorated the cost or inconvenience.[4] Discussions are ongoing about refocusing public breeding to do only basic research – with further development and commercialisation by the private sector (Beingessner 2004b; Carew 2000). While this would certainly benefit the private sector, as basic research takes longer and costs more than the final stages, it would negatively affect breeders by narrowing their activities and making them dependent upon corporate agendas.

IPR can also exacerbate limited diversity within breeding materials, though not simply. Diversity in seeds (and other propagating material) is the foundation of breeding, and limited varietal and genetic diversity have proven problematic at several moments in history, and yet uniformity continues to be the norm. Louwaars et al. (2009) explain some of the complexities. On the one hand, new breeding methods make it faster and easier to use more diverse genetic materials accessed from genebanks, which suggests increased genetic diversity is not only desirable but more easily achieved than in earlier lab-based breeding efforts. On the other hand, however, corporatisation – of materials and methods – and trends toward 'trait breeding' point to decreasing genetic diversity. Moreover, with

---

4   With decreasing government allocations, breeding costs are being downloaded onto growers. Farmer check-off programs, for instance, in which farmers pay a percentage of their crop sales to growers' organisations may provide support to partially fill gaps in funding and breeding.

corporate concentration and increased control of research agendas, the diversity of commercial varieties will likely continue to decrease.

In sum, breeders may benefit from IPR, through licensing fees for instance, but their future options are also inhibited. Limits on material access and diversity coincide with corporate seed ordering processes, constraining research, particularly by public breeders. The socio-economic impacts of IPR though, are not confined only to breeding: first, because restrictions in breeding impact wider agrifood possibilities for researchers, growers and eaters; and second, because growers and savers are changed in their own ways.

Growing and Saving with Intellectual Property Rights

Growers are also concerned with seed-related IPR in Canada, and indicate that even the lower levels of protection provided by PBR are not particularly beneficial for them. One industry-commissioned survey of large-scale grain farmers indicates that only 15 per cent of farmers acknowledge any benefit of PBR, citing higher seed prices, less control, inability to use or sell their own seed, restricted purchasing options, and limited seed supplies as negative implications of the PBR Act on their activities (Blacksheep Strategy Inc. 2004). In addition, many of these surveyed farmers were unaware of the PBR Act or how it affected them (Blacksheep Strategy Inc. 2004), indicating not only limited enforcement – as noted by industry – but also lack of relevance to most farmers' lived experience. Of those farmers who were aware of the Act, approximately two thirds said that because of it they were less likely to sell their crop for seed, although another 25 per cent said they would be just as likely to do so (Blacksheep Strategy Inc. 2004). These data suggest that while some grain growers retain their commitment to seed trading and saving as part of farming practices, the PBR Act constrains these activities if in a limited way.

In addition to more practical concerns already mentioned (here and in the previous chapter) – including increased costs and limited suitable varieties – the proposed changes to the PBR Act suggest a basic devaluing of seed saving and of those involved. Whatmore (2002) draws attention to distinctions made with IPR between invention and heritage – between singular intellectual creativity and collective practiced knowledge. She argues: 'IPR combine the universalizing pretensions of science and law to affect a radical break with the past, collapsing botanical becomings into the here and now of invention such that a germplasm without history is folded into a future of monopoly entitlement' (2002: 110). The tie Whatmore makes here among savers and seeds and their altered possibilities in relation to IPR is at the base of claims for farmers' rights (as discussed in the next section). In places where small-scale growing is the norm, savers continue developing seed, and saved seed plays a fundamental role in survival. As an evocative comparison, farmers are estimated to breed and adapt more than a million varieties every year, while UPOV figures show titles attributed to 10,000 or so varieties a year on average in recent years (Dutfield 2011). In Canada, saving is less practiced and growers less reliant on saved seed, though it does depend on the sector and scale in question (wheat growers save more of their seed than do canola growers, for instance). But what is of concern in any context of PBR or patents is a shift to devaluing seed saving and the complex lives and associations of seeds.

The devaluation of saving practices is most obvious in the case of patents, which do not offer any kind of exemption for growers. Seed saving here disappears as an option, reappearing only as violation – infringement on owners' rights. This revaluation is also increasingly present in PBR, as demonstrated by UPOV91's move to farmers' privilege. This is not just semantics. By requiring an exemption for growers, the revised PBRs of UPOV91 legitimate laboratory science breeding and set it in opposition to the practices of savers. Further, by labelling seed saving activities, and very restricted ones at that, a privilege, extensive PBRs become the norm within which growers may be allowed to practice saving. This radical break is, therefore, normalised in legislation, again displacing seed saving. In the scenarios offered by patents and PBR growers are progressively transformed into potential criminals as they engage in age-old practices of saving (protected) seeds.

For growers, IPR raise concerns that are practical as well as ethico-political. IPR change how (and whether) growers continue to do what they do as savers, as well as revaluing saving, savers, knowledges and seeds.

Making Public Policy?

Of all the responses I heard and overheard at Seedy Saturdays in 2005, the most common response was dismay about the measures, and anger at the lack of public engagement. In the workshops few people were clear on the proposal to shift Canada to UPOV91. As we spent time working these out, people shared their concerns about making such public policy. First, there was a strong shared view that government needed to publicise its proposals more widely and effectively. People felt blind-sided by the proposals and the lack of information, and worried that they might not have heard about the policy directions being undertaken despite the significance of the proposals.

Second, there was a strong negative response to the possibility of further restrictions or even prohibitions on seed saving practices (including exchanging and selling saved seeds). And distress with the discourse of 'farmers' privilege' was expressed at two of the three workshops on policy I facilitated at Seedy Saturdays in 2005. Individual comments against this characterisation of seed saving as abnormal received widespread positive reinforcement – nods and vocal agreements – around the rooms. Some participants expressed disbelief that governments (or industry bodies) would enforce restrictions or monitor seed saving at smaller scales, and therefore their own practices would likely be unaffected. For most, however, this was not an excuse for inaction, instead their concerns about limited seed saving extended to those who would be affected by proposed changes, and to future possibilities of further restrictions. Almost all expressed their intent to contact political representatives to register their disapproval of the proposals, and the lack of publicity.

People also insisted that governments should not just inform but actually engage the public in democratic discussions about seeds and seed saving. In these public discussions, suggested participants, priorities and values should be established *before* policy proposals are made, and government should ensure the public's views are reflected within legislation. Many of the political decisions related to seed-people relations are considered by governments – including Canada's – to be technical matters, and as such, the underlying beliefs and priorities need not be discussed. In its own analysis of public consultations on seed-related regulation, the CFIA indicates that public concerns have been expressed about seed policies and links to globalisation, corporate control, genetic engineering, and more. However, these issues are considered by the CFIA to be beyond particular regulation proposals (CFIA 2007). Participants in the workshops and feedback on the government's consultation, however, indicate that instrumental and consultative approaches are inadequate.

In sum, plant breeders' rights, especially if revised to UPOV91, and patents push seed saving restrictions further, in some cases criminalising what is common practice for many growers. Grower groups, non-governmental organisations and members of the public resisting proposed changes to the PBR Act do so, in part, on the basis that the changes do not reflect growers' needs but instead increase their burdens and limit their options – and opportunities for agrifood ordering. How to ensure seed saving as foundational for food and agriculture, rather than anachronistic or exceptional, is an ongoing debate (c.f. BEDE and RSP 2011; Kloppenburg 2010; Kneen 2006; Zerbe 2007). But it is at least evident that the relations of IPR with seed saving require more attention in the interests of breeders, growers and eaters – whether in the global South or the global North. Increasing worries about IPR, their impacts and their relations gave rise to the idea of 'farmers' rights', a subject that the next section explores.

Encoding Farmers' Rights

Growers have long practiced seed selection, storage and exchange; however, farmers' rights as a policy concept is quite new. It took shape in the 1970s and 80s as international discourse increasingly revolved around commercial property rights, and advocates of other ways of understanding seeds, growers and their relations became even less heard (Borowiak 2004). The proposal for farmers' rights was made within the Food and Agriculture Organization (FAO) by non-governmental organisations (RAFI, now ETC Group, and GRAIN)[5] in 1986. These farmers' rights included saving, exchanging and selling practices in combination with broader socio-economic justice concerns – such as rights to land, agricultural research, well-being and knowledges. Farmers' rights in this form were intended to challenge seed-related intellectual property rights while recognising the contributions of growers to crop diversity (Andersen 2005). To this end, advocates argued that growers' practices needed to be recognised, legitimated and supported in ways that moved beyond the granting of an exception to commercial rights. Farmers' rights have since proven provocative; in their various forms they have been strongly supported and/or protested within international forums, national policies and non-governmental advocacy (Anderson and Winge 2011; Brush 2007; Feyissa 2006a; Zerbe 2007).

Since its appearance within international negotiations, the concept of farmers' rights has been interpreted in diverse ways. In her discussion of the concept's history, Andersen (2005: 23–4) suggests that within international negotiations farmers' rights have included four themes: balancing breeders' rights; rewarding growers as a community; conserving biodiversity and related knowledge; and establishing an international fund. By comparison, La Via Campesina, a worldwide umbrella peasant organisation arguing for food sovereignty, has a more detailed view. In a submission to the FAO in 1996 pleading for the recognition of farmers' rights, this organisation listed a series of points relating to seeds including local communities' rights to: control seeds and decide their futures; participate in defining and executing related policies; define the legal framework applied and benefit-sharing arrangements; develop sustainable agricultural methods and influence supportive policy; and, freely use, choose, store and exchange seed. More generally, Correa (2000) identifies equity, conservation and preservation of farmers' practices as the primary rationales for farmers' rights, and although both Andersen's and La Via Campesina's versions comply with Correa's framing, each offers variations on themes of farmers' rights and why these rights are significant.

In international agreements acknowledgement of farmers' rights and efforts to support them manifest distinctly, depending on differential power relations within each institution and its negotiations at that time. As noted earlier, UPOV91 provides for optional and limited farmers' privileges, while TRIPs does not recognise farmers' rights at all and its *sui generis* clause is disputed as providing an avenue for recognising farmers' rights. Two additional agreements warrant mention: the Convention on Biological Diversity (CBD) and the International Treaty on Plant Genetic Resources for Food and Agriculture (ITPGRFA, 'the Treaty'). First, the text of the CBD makes no mention of farmers' rights

---

5   Both of these NGOs are international in focus and in structure; however, each has Canadian presence.

in particular, but it does contain provisions to respect, preserve and maintain traditional knowledge, innovation and practices (subject to national legislation) that some have interpreted as providing room for debate around farmers' rights. Further, the 2006 eighth CBD Conference of the Parties urges members to 'respect traditional knowledge and Farmers' Rights to the preservation of seeds under traditional cultivation.' This is a limited acknowledgement, and one with no practical follow-through, but it does provide some recognition of the notion of farmers' rights. The second relevant international agreement, the ITPGRFA, contains specific provisions on farmers' rights. It is considered to be the most hopeful in relation to farmers' rights inclusions, and these provisions orient the next subsection.

Farmers' Rights in the ITPGRFA

The FAO's International Undertaking on Plant Genetic Resources and its successor, the legally binding ITPGRFA, offer the strongest endorsement of farmers' rights within international agreements. In 1989 resolution 5/89 of the FAO's International Undertaking on Plant Genetic Resources strongly endorsed farmers' rights as 'arising from the past, present and future contributions of farmers in conserving, improving, and making available plant genetic resources, particularly those in centres of origin/diversity.' It was a major accomplishment for farmers' rights advocates, though it came with the concession that PBR did not conflict with the Undertaking. Despite the strong endorsement however, the operationalisation and conceptualisation of farmers' rights was limited since the Undertaking was not legally binding, did not include implementation measures, and did not provide a clear definition. The International Undertaking has been superseded by the ITPGRFA, a Treaty spurred by inconsistencies between the Undertaking and other international agreements, as well as recognition that 'plant genetic resources for food and agriculture' (see Chapter 5) required particular attention.

The Treaty, which entered into force in 2004, acknowledges farmers' rights in two sections. First, Article 5.1.c. directs member states, subject to national legislation, to 'promote or support, as appropriate, farmers and local communities' efforts to manage and conserve on-farm their plant genetic resources for food and agriculture'. Second, Article 9 more directly acknowledges farmers' rights by recognising the 'enormous contribution' of indigenous communities and farmers in conserving and developing the basis of food and agricultural production. Article 9.2 stipulates:

> The Contracting Parties agree that the responsibility for realizing Farmers' Rights, as they relate to plant genetic resources for food and agriculture, rests with national governments. In accordance with their needs and priorities, each

Contracting Party should, as appropriate, and subject to national legislation, take measures to protect and promote Farmers' Rights, including:

a) protection of traditional knowledge relevant to plant genetic resources for food and agriculture;

b) the right to equitably participate in sharing benefits arising from the utilization of plant genetic resources for food and agriculture; and

c) the right to participate in making decisions, at the national level, on matters related to the conservation and sustainable use of plant genetic resources for food and agriculture.

Article 9.3 further clarifies that nothing in the Treaty should be interpreted as limiting 'any rights that farmers have to save, use, exchange, and sell farm-saved seed/propagating material, subject to national law and as appropriate.' While farmers' rights are recognised, and to some degree defined in this text, states remain responsible for the legislation and realisation of these rights. Further, the meaning of 'as appropriate' remains open. With the inclusion of these two conditions, the Treaty's version of farmers' rights remains largely a moral plea rather than comprising concrete measures.

In its text then, the FAO's ITPGRFA, which Canada has joined, includes weaker support for farmers' rights than did the International Undertaking; however, it is still the strongest endorsement of farmers' rights in international agreements. The stance in favour of farmers' rights by the Treaty would likely have been stronger had not some state representatives, including Canada's, intervened during negotiations to diminish farmers' knowledge while lauding techno-scientific expertise and innovation (Whatmore 2002: 112). However, some civil society organisations are working hard to build upon this moral imperative, pushing their agenda for a broader understanding of farmers' rights, and progress is being made.

Limits and Possibilities of Farmers' Rights

Farmers' rights, as collective and dispersed rights, are difficult to conceptualise and operationalise within the contemporary legal IPR framework, which relies upon and reinforces private, individualised property rights. However, farmers' rights have been criticised for more than their limited conceptualisation and operationalisation, even by those who advocate the resistance of IPR and their effects on growers and eaters. The main concern is that farmers' rights may serve to legitimate the very IPR regime that they emerged to challenge in two ways: by evading direct confrontation with IPR; and by employing rights language.

First, farmers' rights do not necessarily confront breeders' rights; rather, they make claims of their own that may be separately addressed (Borowiak 2004;

Brush 2007). In this way, breeders' rights, patents, and their related systems can continue to become stronger, while farmers' rights, even if recognised, remain ill-defined and/or poorly implemented. Acknowledgement and endorsement of farmers' rights in international agreements such as the ITPGRFA, for instance, may give the impression that growers are protected and that there is some level of equity between breeders and growers, whereas in fact this is not the case. The same applies at the national level. As discussed in the case of Canada, the proposed farmers' privilege clause would actually serve to further enclose seed saving practices, while appearing to offer growers legal protection for the first time.

Second, by using the discourse of rights, advocates of farmers' rights implicitly (if not explicitly) legitimate the idea that rights are a valid means of discussing and ordering seed-people relations. The use of rights discourse can prevent discussion of other options for resistance, including those based in broader understandings of social justice or citizenship. As Kneen (2006) suggests, employing a discourse of rights implicitly legitimates governmental power over seeds and growers by (falsely) suggesting that the practice of saving seed cannot exist without state permission. Further, farmers' rights discourse does not question underlying issues with the evolving seed order including the treatment of seeds as commodities, the status of corporations as legitimate private property claimants, or the power relations among the variously involved agents. Farhad Mazhar (2007: 15), a leader within Nayakrishi Andolon (New Agricultural Movement) in Bangladesh, makes the importance of power relations, rather than rights protection, clear when talking about the saving of seed as a claim of power; he states that 'what we are indeed discussing is a battle, not for "rights" or "property", but for power, a battle between corporations and the people of the world.'

Although the concept and implementation of farmers' rights is limited in several ways, it is suggestive of resistance to seeds becoming privatised commodities and savers becoming potential criminals. First, recognition of the co-evolution of people and seeds and of farmers' rights within the Treaty, even if ineffective in practice, does indicate some unease in international arenas with increasingly dominant understandings and approaches to seeds, growers and their interrelations. Acknowledging growers and seeds as co-evolving unsettles a strict division between nature and culture upon which the dominant perspective relies. Primarily concerned with inequitable socio-economic ordering, farmers' rights remain human-centred (endorsing a kind of stewardship of seeds by growers). However, seeds regain some of their histories and broader valuations by virtue of their connections with growers. In this recognition of the relations joining seeds and savers, farmers' rights partially disrupts the earlier discussed differences made by IPR.

In addition, farmers' rights, even in limited forms, contain elements that imply challenge to the dominant private property version of IPR (Borowiak 2004;

Patel 2007). Part of the difficulty of defining and implementing farmers' rights is a reflection of its contestations about what rights should be – collective or individual, enacted or legal, tied to responsibilities or not, state-based or otherwise, and so on. Farmers' rights might be used to challenge how rights are composed, and resulting relations, spurring a broader questioning of dominating and durable agrifood orders and their composite values, discourses and practices.

The possibilities and controversies of farmers' rights and the vagaries of the ITPGRFA have spurred efforts of clarification and realisation. The Fridtjof Nansen Institute in Norway, for instance, has been engaged in a long-term project to research and support farmers' rights, building on the recognition within the Treaty (Farmers' Rights 2012). Developed case studies, surveys of 31 countries and a global consultation are just part of what the project has involved. In the Institute's global consultation, for instance, recommendations of regional groups were collected, compiled and presented to the fourth session of the governing body of the Treaty held in 2011 (see Andersen and Winge 2011). Further, recent moves to build upon the food sovereignty movement and include farmers' rights under a framework of seed sovereignty promise both contest and reconstitution of seed orders.[6] Though food and seed sovereignty have their own conceptual and programmatic challenges (see Patel 2010), moving farmers' rights toward seed sovereignty provides opportunity to regain the more robust, politicising notion originally intended.

Seed sovereignty is beginning to find footholds. Community, regional and national seed networks are growing in numbers and in influence, and increasingly use seed sovereignty as part of their discourse (see Navdanya 2012). Further, international organisations such as the European Coordination of Farmers' Seeds and La Via Campesina are using the ITPGRFA to demand that farmers' rights be included in legislation and programs of member states. The Bali Seed Declaration, a 2011 civil society statement presented at the fourth session of the Treaty, makes the drive for farmers' rights within a wider food sovereignty agenda clear:

> We cannot conserve biodiversity and feed the world while our rights to save, use, exchange and sell our seeds are criminalized by laws that legalize the privatization and commodification of seeds. … We demand that farmers' rights be mandatory and that the rights of breeders be subordinated to these farmers' rights.

During this latest session, members adopted Resolution 6/2011 requesting the convening of regional workshops on farmers' rights, encouraging the development of best practices of implementation, and inviting financial and

---

6   Food sovereignty as a concept was initiated by La Via Campesina in 1996. For academic explorations of food sovereignty, in some cases as applied in global North contexts, see Wittman et al. (2010; 2010); Patel (2009).

technical support for implementation at national scales. There is, therefore, opportunity to take advantage and further open the spaces made by the Treaty to advance seed saving practices.

Farmers' Rights in Canada?

In Canada, farmers' rights remain in question. Current legislation, reflective of UPOV78, does not refer to farmers' rights in particular; however, the conditions of plant breeders' rights allow growers to save and exchange protected seeds as long as these practices remain non-commercial. Although the existing PBR Act serves to contain growers' practices by excluding commercial activities, the non-commercial saving and exchanging of protected seed remains permissible by virtue of its omission. In the CFIA's proposed changes to the PBR Act, which reflect UPOV91, farmers' rights are still absent, but not in a facilitating way. Instead, farmers' privilege is offered as a possible exception to PBR such that (if adopted) the exception would allow farmers to save seed for reuse on their own lands for their own purposes. Not adopting the clause within a changed Act would mean growers would lose any rights to saving seed from protected varieties, while adopting the exemption further constrains seed saving such that any exchange, commercial or otherwise, would be disallowed during the period of protection.

The fact that farmers' rights are not articulated within Canada's current legislation has led to arguments over whether farmers' rights actually exist. There is no dispute that growers have saved and exchanged seeds and continue to do so (and that governments and corporations have been aware of these practices). However, whether this practical reality suggests the existence and recognition of farmers' rights per se has been questioned. Since the term is not legally encoded, some industry representatives argue that farmers' rights have never existed; instead, there has simply been a lack of prohibition of the long-standing practice of growers saving and exchanging seeds (CSTA lawyer as quoted in Wells 2005). The issue then becomes not that proposed changes prohibit (or further restrict) the practice of saving seeds, but that the legally encoded plant breeders' rights need to be protected from infringement by savers. In this way, savers become not defenders of rights, but infringers upon legal rights and claimants of false rights. Of course, growers have articulated another perspective entirely, maintaining that the traditional practice of saving seeds (including exchange practices) is part of common culture and should be legally protected even if the practice has not been encoded as farmers' rights per se.

At least some of the Canadian public agrees that policy-making should engage with seed saving in more facilitative ways. In 2004–2005 over 35,000 signatures were gathered on petitions submitted to Parliament, opposing the

CFIA's proposed changes to the PBR Act and demanding acknowledgement of farmers' rights. Various public talks and protests were held across the country. In a presentation to organic growers, Dan Jason (2005) – seed saver, small seed company owner, and seed sanctuary facilitator – articulated the surprise and hope that emerged for him, among others, with these efforts:

> I want to say that this seed issue has captured so many people's imaginations and disgust across Canada. And, we downloaded the NFU petition, just, well, it must have been about two months ago. And we had it in all our stores in Salt Spring Island. It was unbelievable the conversations that were going on in Thrifty's [a supermarket][group laughter] ... just everyone outraged that this kind of thing is going on now. And I think it's time to really just jump on this and move in a totally positive direction for what we could be doing in Canada in terms of maintaining our whole heritage of all the stuff that's been passed on forever and ever and ever.

As he tells it, as Dan was going about his mundane grocery shopping, he became part of a shared political engagement that has inspired him not only to resist but also to seize this moment of critique to move Canada's seed networks in a more positive direction.

Later that year Gail, a hobby grower and serious amateur seed-saver, shared with me some of the fear that came with proposed policy changes, the Schmeiser verdict, and increasing corporate involvement with seeds, saying:

> And what I've been a little bit afraid of, with all the legislation and that coming, is kind of, there's a feeling that some of us don't want to be on the radar because if push comes to shove and they start enforcing any legislation or regulations, then if you're on the radar then of course you're part of it. And quite frankly, I'm just speaking for myself [and my family], we don't want to be part of it.

Gail's fear of legislation and prosecution conflicts with her dedication to saving seeds; the fear and a desire to protect herself and her family advances a sense of wanting to stay off the radar, but she is directed otherwise by her commitment to seed saving, building her community's seed network, and sharing information about seed-related policies and issues. Gail, and a few of the other savers I spoke with, experience tension between a sense of security from being unseen and desires to increase and improve seed saving efforts. Fear of this kind might propel people to hide their practices, disengage from wider debates or community relations, or even stop saving seed. But for those with whom I spoke, often their resolution was just to get on with seed saving – while keeping an eye on wider issues.

One motivation for many savers is resisting efforts – governmental and corporate – to shut down seed saving and sharing. Chapter 8 more deeply explores the idea of seed saving as political engagement, but here I want to mention one seed saver's articulation of how being confronted by IPR can be motivating. Rachel shares this sentiment by relaying her experience with PBR labelling on a plant she purchased from a local nursery. She recounts:

> There was this huge, bigger than the information about the plant, was the warning, about propagating Hebe. ... we [her friend and she] joked about it for a really long time, about illegal Hebe propagation and that we were going to start a Hebe farm. And how would they know? Were there little sensors attached to the Hebe? Were there going to be Hebe police? You know, what's going on? And suddenly I want to! I think I'd like to make a lot of Hebe, you know?

This sense of rebelling against what is considered illogical and inappropriate legislation enclosing saving and the lives of seed and savers is shared by many of the seed savers with whom I spoke, whether as part of research or casual conversation. Thus far, saving practices persist – with or without encoding as farmers' rights – despite being threatened by stronger IPR.

Summing Up

Seedy Saturdays this past year (2012) had a more relaxed feel to them than did those of 2005 – at least they seemed that way to me. After public protest, the proposed changes to the PBR Act were put on the backburner. Their re-appearance is expected, however, most likely in the form of a trade agreement negotiated in backrooms away from public scrutiny. Since the time of the proposed changes, seeds and saving have garnered more attention internationally and nationally as part of alternative food network efforts, but also in campaigns dealing particularly with seed saving. While the increased awareness of seed issues and experiences of other countries may mitigate seed savers' shock at their government pursuing measures to contain seed saving, the concerns about the proposed changes – their wrong-headedness, the lack of public engagement, problematic language – still apply. Savers are more aware and wary now – as are governments.

These are not only domestic concerns; Canada's shift toward stronger property rights protections coincides with international trends in multilateral and bilateral agreements. The practical impacts of IPR for breeders and growers such as decreased research, limited seed variety and restrictions on saving are concerning in their own right, but the implications of IPR go further. Seed-related IPR assert and authorise particular boundaries and entitlements in ways that reinforce and advance socio-economic and ethico-political inequalities. IPR further the corporate reordering of seed discussed in Chapter 3 by enabling privatisation of seed and breeding while legally limiting seed saving practices. However, characterisations of the provisions in these agreements as self-evident and universal are problematic.

There are gaps in which resistances find footings, and evasions persist. Several examples of contestations of IPR have been given in this chapter – including court cases, public policy advocacy, and even individuals balking at unreasonable policies. The concept of farmers' rights has been particularly explored as a potential counter to seed-related intellectual property rights. Like other engagements, farmers' rights has its own limitations, but, particularly when combined with a wider agenda for food sovereignty, it opens possibilities to critically engage and challenge IPR, their impacts, and their furthering of private capital's seed reorderings.

Chapter 5
Securing Accessions

Almost 800,000 samples. Over 500,000,000 seeds. I can't imagine it.

This is what the Svalbard Global Seed Vault (SGSV, the Vault) currently stores (NordGen 2012a). Apparently, the Vault's capacity is three million samples, and it represents the most diverse collection of food seeds in the world (GCDT 2011, 2012a). Each sample varies depending on seed bodies and depositor's preferences, but there must be, according to guidelines, at least enough for two successive regenerations – usually around 500 seeds (NordGen nd).

The seeds are stored in foil packages, in black plastic boxes, in three rooms, set behind three successive sets of doors, at the end of a 125-metre tunnel into a mountain, on a remote set of Arctic islands in Norway (GCDT 2012a). All genebanks store seeds *ex situ* – 'out of place' or removed from their growing contexts – but the Vault offers another reality of *ex situ*.

Dubbed 'the doomsday vault', the SGSV opened to plenty of media attention in 2008. It stores duplicate seed samples from collections around the world with the aim of ensuring, in perpetuity, the conservation and availability of crop diversity for global food security (GCDT 2012a). The idea for such a facility evolved in the 1980s as part of the advocacy for conserving rapidly-disappearing crop diversity, but only with the ITPGRFA (the Treaty) did it become a real possibility (Fowler 2008; Mooney 2006b). The SGSV is jointly managed by the government of Norway, the Nordic Genetic Resource Centre, and the Global Crop Diversity Trust (GCDT, the Trust) – an international organisation established within the Treaty framework and funded by a variety of governments, corporations and organisations (FAO 2006, GCDT 2011).

The SGSV prioritises seed of 64 food and forage crops listed in the Treaty (see Appendix 3), with some attention to minor crops and wild relatives (Fowler 2008).[1] The natural permafrost conditions are supplemented to ensure temperatures of -18 Celsius, and humidity is kept low – meeting international standards for long-term seed storage. Some of the priority crops – like breadfruit, potato, and banana – cannot be effectively conserved as seed, reproducing best through vegetative means, so at other sites cryopreservation methods keep plant material using liquid nitrogen (GCDT 2011).

---

1   Although the GCDT remains open to the option of storing GE seeds in the SGSV, at this point Norwegian law does not allow it.

Accessions – deposited seed samples – come from national genebanks such as Canada's Plant Gene Resources Centre, the international research centre network of the Consultative Group on International Agricultural Research (CGIAR), and even saver organisations such as the Seed Savers' Exchange of the United States (NordGen 2012b). These seed depositors are responsible for the testing and regeneration of vaulted seeds (NordGen nd), but recognition that many genebanks – particularly in developing countries – are unable to manage duplicating their collections has led the Trust to assist in pre-deposit regeneration and preparation (GCDT 2012b).

The depositor agreement also requires that any seed samples in the Vault be made available to all by depositors from their own collections (NordGen nd). This means vaulted seed remains inaccessible, but seeds become accessible through depositors' collections.[2] Accession information is being made available as part of GENESYS – an online portal connecting vault information with other existing international genebanking information databases (NordGen 2012).

The tale of the SGSV offers clues into contemporary technopolitical arrangements for keeping seed, and human agrifood futures, secure. The Vault is an extreme example, a 'doomsday' backup, but its form is based in knowing and valuing seeds in ways common to genebanking practices. As the Vault illustrates, a genebank is a structure that houses accessions, but, more than this, it assembles scientists and seeds, cooling rooms and digital networks, conservation and use values, funders and contract agreements, and more besides. Genebanking is a type of *ex situ* conservation in which 'plant genetic resources for food and agriculture' (PGRFA) are stored, scientifically assessed and catalogued, and made ready for research and breeding purposes. These practices continue long-standing efforts of maintaining and exchanging useful seed (aka seed saving) and efforts to collect seed as part of political strategies, but do so in their own way.

This chapter takes genebanking – its practices and arrangements, attachments and challenges – as its subject. The first section outlines the progression toward genebanking, considering genebanking within a longer tradition of collecting useful seed/plants. I then explore contemporary genebanking, examining current collections, enactments and practical challenges. The final section reveals some of the underlying issues related to genebanking, focusing on ways in which genebanking may: reinforce existing socio-economic inequities, specifically between breeders and savers as well as between global North and South countries; reconstitute seeds as accessions of PGRFA in ways conducive to commercial interests; and perpetuate genetic reductionist approaches to knowing seeds. I argue that genebanking serves and reinforces the corporatising

---

2   There has been controversy over the Seed Savers Exchange's decision to use the SGSV specifically in relation to this aspect of Depositor's Agreement (see Whealy 2010; CFS 2011).

seed order – in ways direct and indirect; however, its relations also suggest other priorities and possibilities.

Becoming Genebankers

Plant collection and control has long participated, and been enrolled, in political strategies, with earliest official accounts coming from ancient Sumerians collecting figs, roses and vines around 2500 BCE (Fowler 1994). Military campaigns, explorers' expeditions, and colonialist expansions relied upon and spread crops, cultivation methods and ways of knowing seeds in the service of particular interests (Brockway 1988; Drayton 2000). Seeds, knowledge about how to grow and use them, and *ex situ* storage – botanic gardens, agricultural research centres, and most recently, genebanks – are fundamental to keeping and extending particular agrifood practices, and power.

In Canada it was not until the late 1800s that the state became interested in seed collection as a means of increasing agricultural productivity. Though there were seed merchants travelling Canada in the early 1800s, it was primarily the waves of early European immigrant farmers and gardeners who developed varieties and distribution networks to serve their needs, learning to sustain themselves anew (Martin 2000; Woodhead 1998). Despite some adoption into settler diets and cultivation, colonialism destroyed much of the biocultural diversity developed over generations by indigenous peoples – a situation several current saver initiatives attempt to redress (Kuyek 2004b; LaDuke and Scott 2011). Only in 1886 did the Canadian government initiate a program, based on the successful model already running in the United States, to collect seeds from around the world, multiply samples at research centres, and then distribute packets to farmers for further testing and selection (Kuyek 2004b; Symko 1999).

Similar efforts in the USSR resulted in one of the most significant seed collections amassed before long-term storage in genebanks. Under the guidance of Nicolay Ivanovitch Vavilov (1887–1943), the primary aim of Russian collection was to provide varieties adaptable to difficult agricultural conditions. In addition though, he included the objective of conserving seed for future food security. Vavilov was one of the first to perceive a threat from loss of agricultural diversity, and he made these dual goals clear upon meeting his staff at the All Union Institute for Applied Botany and New Crops in 1923 (Fowler and Mooney 1990: 148). Passion for scientific knowledge, food security and exploration drove Vavilov, and he and his colleagues collected seeds within the USSR and over 50 other countries (Nabhan 2009).[3] By 1940 around 200,000

---

3   Based on these expeditions and subsequent analysis, Vavilov believed that there were several 'centres of origin' around the world. His thesis has since undergone multiple revisions, but Vavilov's insights still inspire.

accessions were stored in the collection (Reznik and Vavilov 1997), and it is this remarkable collection that serves as the basis of Russia's current genebank – now known as the NI Vavilov Institute of Plant Industry.

As these seed collections were growing, ways of developing and knowing seeds shifted, changing practices of collection, distribution, storage and breeding as well as involved practitioners. In the early 1900s increasing political and commercial interest in industrialised agriculture coupled with the rediscovery and further development of Mendel's laws of inheritance (which outline seed trait continuance over generations) propelled a dual move. First, laboratory breeding by scientific experts displaced in-field selection by farmers and breeders, and second, a focus on identifying inheritable traits that could be bred into seeds replaced efforts to gradually adapt and develop seed varieties (Kloppenburg 2004: 68–9; Kuyek 2004b).

Lab-based processes concentrating on genetics allowed specialists to choose advantageous traits and eliminate unwanted ones. With new insight and methods, plant scientists began developing 'pure lines' – stable, distinct, uniform cultivars (Fowler and Mooney 1990). Further, by breeding two desirable pure lines together, they created marketable 'improved' hybrids.[4] Hybrid corn developed this way in the 1930s and proved wildly successful, with other crops following in its wake (Kloppenburg 2004). The seed collections gathered from farmers' fields around the world that once provided seed to be adapted became sources of genetic material with useful traits, and they adjusted with the times.

In some ways, particularly in relation to yield, improved varieties demonstrated superiority, but their widespread adoption came with obligations and losses. These seeds needed to be purchased year-to-year to maintain their improved traits. Further, these seeds carried with them agricultural systems favouring chemical inputs, monocultures and industrial technologies. As well as making choices, growers adjusted to these seeds. During the Green Revolution of the 1960s and 1970s these seeds, already widely adopted in the global North, were promoted in the global South by governments, industry and international development agencies as part of farming packages aimed at modernising agriculture.

As part of the drive to spread these new seeds and their accompanying agricultural practices, International Agricultural Research Centres (IARCs, now part of the CGIAR network) were established in areas of crop diversity in the global South. The Rockefeller and Ford Foundations, in collaboration with US and hosting governments, created the Centres, extending existing seed collection programs. In addition to promoting commercial hybrids in

---

4   In hybridisation two plant lines with promising traits undergo isolated self-pollination over multiple generations until a pure line is achieved. These pure lines are then crossbred resulting in an F1 hybrid. F1 hybrids' next generations express desired traits erratically ('reverting' to parent lines), making repurchase necessary to retain full benefits.

their regions, the IARCs undertook significant collecting and classifying of regional crops, later providing the foundation for FAO genebanking pursuits (Pistorius 1997: 5–7). These 'new and improved' seeds, therefore, were deeply implicated in complex processes of changing agrifood orders, and agricultural diversity suffered.

While savers have always selected seeds that suit their various needs, and sometimes this means replacing particular varieties, the loss that came with the improved, uniform cultivars is remarkable. It is estimated that 75 per cent of agricultural diversity has been lost since the 1900s, with over 90 per cent of crop varieties having disappeared from growers' fields (FAO 2004b). Already lost to improved varieties and altered agricultural practices: 7,000 rice varieties of Bangladesh; 80 per cent of the maize in Mexico; almost all of the sorghum in South Africa; and 90 per cent of the 10,000 varieties of wheat grown in China (GCDT 2002: 6; Nazarea 2005: 6; Thrupp 2000: 269). The list could go on and on. The loss of in-field diversity continues today, with ongoing adoption of hybrids and accompanying agricultural practices, but also with promotions of GE seeds (sometimes referred to as the Gene Revolution).

Much of the power of arguments for conserving crop diversity comes from the threat that accompanies lost diversity. Warnings about the effects of uniform, hybrid varieties on agricultural diversity were common in the 1960s and 1970s. Similar concerns were raised decades earlier – as early as 1936 by US plant scientists Harlan and Martini – giving rise in the 1940s to periodic national and international deliberations (Fowler and Mooney 1990: 61). It has been widely acknowledged for some time that limited genetic and varietal diversity leads to increased vulnerability, especially in relation to people's food security (Bellon 2003; Brush 2000; Harlan 1975; Lammerts et al. 1999; Thrupp 1998). Further, as seeds disappear so too does knowledge of their histories and of how to grow, save, prepare and consume them (see Nazarea 1998, 2005; Pfeiffer et al. 2006; Wertz 2005). Potential loss is seen as threatening to human futures, raising questions like: what if a plant disease becomes rampant and food supplies dwindle?; what if growers favour varieties that are less genetically diverse?; or what if savers select out potentially useful traits?

Times when the unspecifiable 'what ifs' emerged as reality lend further support to fears about the possible implications of lost crop diversity. Spectres of the Irish potato blight of the 1840s, the loss of coffee production from parts of Asia and Africa in the 1870s, the US corn crop failure in 1970,[5] among others, point to the possible tragedy. In each of these cases, the resistance required was found in varieties developed in-field by savers – illustrating the importance of retaining and developing diversity. Varietal diversity is relatively easily assessed

---

5   Notably, later discovery revealed that the responsible fungus infected the hybrids' t-cytoplasm, introduced in hybrid corn varieties to simplify breeding.

and without it genetic diversity is inherently limited, but even with recognition of the threat diversity remains limited in commercial agriculture. As example, all commercial sunflowers grown in the US are hybrids using the same male-sterile cytoplasm despite recognition of crop vulnerability and existence of numerous wild populations within the country (USDA 2007). Genetic diversity, for its part, is often invisible and must be assessed through laboratory tests, making the 'what if' questions and their associated threats, at least partially, left for experts to debate and resolve.[6]

Shaping and Prioritising Genebanking

Despite the ongoing warnings, losses and discussions in national and international forums, it was not until the FAO's International Biological Programme Technical Conference on the Exploration, Utilisation, and Conservation of Plant Genetic Resources of 1967 provided impetus for global recognition of the problem that genebanks emerged as preferred strategies combining the making available for use and the conserving PGRFA.

Deliberations of *in situ* conservation – maintaining crop diversity 'in place' (or in growers' fields through seed saving) – had occurred and received some support. Saving in-field keeps those varieties proven useful to growers going, offers resilience when diverse seeds are used, provides some measure of independence for growers, and fosters new crop diversity through adapting and improving practices (Brush 2000; Engels 2001). But the changing relations of seeds and savers *in situ* came with risk of lost varieties and genetics, and contradicted breeding approaches in which 'genetic resources that remain in farmers' fields are not directly useful for crop improvement' (Brush 2004: 197).

Genebanks, in contrast, offer convenient access to useful material for lab-based breeders to serve commercial, industrial agriculture. Further, genebanking, theoretically at least, saves humanity from the 'what if' threats by ensuring that potentially useful traits are preserved in secure locations for later use. Through this view, the flexibility and adaptability of seeds in connection with savers, once understood as strengths in the changing circumstances of growing foods in the fields, become seen as problems. Genebanking, therefore, becomes necessary to deal with insecurity in-field. Ironically, it does so even as it reinforces the practices that gave rise to those losses and threats. At the time, in debates over how best to maintain plant genetic resources, genebanks won.

In 1974 the International Board for Plant Genetic Resources (IBPGR, later International Plant Genetic Resources Institute (IPGRI), and now Biodiversity

---

6   Although diversity in phenotypic traits can indicate genetic diversity, they are not the same thing; genetic diversity may exist without phenotypic expression, and vice versa. Most seed savers I spoke with had not examined seeds' genetic diversity, although many had read information in this regard.

International) was established within the FAO to facilitate global efforts to maintain crop diversity. It was a compromise position: the Board would be housed with the FAO, but the genebanking network – originally intended to be created as part of coordinated efforts – relied upon and extended existing CGIAR IARC collections and national genebanking (Kloppenburg 2004), each with ongoing mandates serving national and commercial interests. With this international stimulus, collecting and genebank construction hastened.

Much was accomplished during the 1970–80s in genebanking crop diversity. Efforts filled gaps in existing collections, developed long-term storage facilities, and improved systems and coordination. By the mid-1980s the FAO's initial goal of 50 genebanks with long-term storage was surpassed, with 75 countries running such facilities (Fowler and Hodgkin 2004). Further, whereas previous collections had focused on usable varieties, genebanks now took on conservation as well.

Both mandates received support and challenge in the 1980–90s from different sectors. The efforts at conserving crop diversity predate the 1980s' explosion of interest in biodiversity conservation, but these concerns still had impact. On the one hand, genebanks benefited from heightened support for conservation, but on the other, conservation interests challenged the more utilitarian focus of genebanking by arguing for wider collection of diversity and prompting a move toward 'sustainable use' (Pistorius 1997: 93–8). In addition, increasing interest in and demand from the ag-biotech sectors drove genebank expansions, this time supporting a focus on useable material and documentation (Kloppenburg 2004). Ag-biotech also released genetically engineered organisms, complicating any genebanking of non-GE seeds. So genebanking received support from new actors, but this support also brought increased and contradictory expectations.

In addition at this time, the formation of non-governmental organisations particularly interested in international seed and agrifood issues provided momentum to debates over intellectual property rights and social injustice of seed arrangements.[7] The IBPGR working with the IARCs did much to conserve quickly disappearing crop diversity, but by favouring storage in the North and in IARCs they also gained control over those seeds (Andersen 2008: 90). Non-governmental organisations, working with global South countries, questioned these inequities and their exacerbation with accelerating applications of IPR. In ensuing controversies over genebanking, seeds were variously figured as biodiversity to be conserved, insurance for future food security, common goods to be shared, commodities to be owned, and resources to be controlled and/or developed. Responses to these debates, and further controversies,

---

7   The Rural Advancement Foundation International (now the ETC Group) and GRAIN each formed during the 1980s. They have since engaged in research and advocacy relating to agrifood issues, including conservation and use of agricultural diversity.

underlie the complex technopolitical environment within which contemporary genebanking operates.

In relation to contemporary PGRFA governance and genebanking, two international agreements are worthy of mention: first, the Convention on Biological Diversity (CBD); and second, the ITPGRFA. The CBD, which came into force in 1993, aims for the conservation and sustainable use of biodiversity as well as the fair and equitable sharing of benefits from use of genetic resources. Two aspects of the CBD are particularly relevant to genebanking. First, the CBD stipulates that each contracting party should establish and maintain *ex situ* conservation facilities and programs, as complementary to *in situ* measures (Article 9). Second, the agreement accepts the principle of national sovereignty over biodiversity, designating decisions about access to states and indicating that mutually agreed terms should be set to share any benefits from PGR research and development (Article 15). Through the CBD then, genebanks gain support for conservation and sustainable use mandates, but bilateral arrangements become necessary for international collection. For some (and in some cases), the CBD provides a mechanism to protect national interests and control PGR, while for others these measures as seen as impeding flow of what should be a common resource (Andersen 2008: ch.6; Fowler 2004; Worede 2006). While the CBD applies to all PGR, crop diversity is dealt with more specifically through the Treaty.

The ITPGRFA, or Treaty, came into force in 2004 as the first legally-binding agreement oriented around 'plant genetic resources for food and agriculture' (PGRFA). The Treaty objectives, in general, are to facilitate access and sustainable use, while promoting benefit sharing (Art. 1). The Second Global Plan of Action for PGRFA reinforces these goals and advances their implementation. In relation to genebanking issues, the chief accomplishment of the Treaty is its multilateral system of exchange for contracting parties in relation to 64 major food and fodder crops (see Appendix 3 for listing), which account for approximately 80 per cent of global caloric intake. This makes bilateral arrangements for accessing these crops unnecessary for contracting members, though crops outside the multilateral system remain subject to CBD measures. To further expand the crops covered, member states are invited to add national collections to the multilateral system, and to encourage private collections to do the same. Through its measures, the Treaty signifies a major shift in PGRFA governance.

Exchanges of PGRFA take place under agreed conditions as laid out in the Treaty and its standard material transfer agreement (SMTA). Among other things, the SMTA stipulates legally-binding conditions for: making PGRFA and accompanying information accessible; application of intellectual property rights; and benefit-sharing conditions and mechanisms. But each measure comes with caveats: most have been made 'subject to national legislation'; accessibility is equated to availability for research; information includes only non-confidential details; and it is on materials 'in the form received' that IPR cannot be applied. Moreover, benefit-sharing is expected only when access to material is limited (i.e., made unavailable to researchers, likely through patents), and it takes diverse direct and indirect forms as determined by the Governing Body. These caveats lessen the impact of more controversial aspects of the Treaty but provide some new options that might be pushed further, much like its provisions on farmers' rights as discussed in Chapter 4.

In sum, genebanking has been entangled with international agreements, national objectives and commercial strategies from its beginnings, and relations between power and seed-plant collection date back even further. Genebanks remain the prioritised method for achieving the dual goals of conserving and making available for use PGRFA, and have become essential, particularly with continuing crop diversity loss in fields, food security threats, and changing climates and agrifood cultures. Today's practices reflect the history of becoming genebankers, disposing genebanks to particular methods and relations. Genebanking continues to alter with current seed reorderings, adapting capacities, activities and methods while working to conserve and make accessible PGRFA. The next section elaborates contemporary genebanking practices and attendant practical challenges, after offering an outline of the current status of genebanks.

Genebanking and Becoming an Accession

Over the past four or five decades *ex situ* conservation has expanded significantly. Worldwide, the number of accessions, or samples, amassed in genebanks and botanic gardens has reached 7.4 million, compared to about half a million in the early 1970s (FAO 2010b; GCDT 2002). Only 25–30 per cent of today's collections are considered unique, with the remainder being duplicates (such as those in the SGSV).[8] National governments control most accessions (90 per cent) and genebanks, with the largest state collections found in China, the US, India and Russia. Worldwide, genebanks number 1,750, of which only 130 hold over 10,000 accessions each. These collections are also concentrating; seven countries now hold over 45 per cent of all state accessions, compared with 12 in 1996. Aside from the SGSV, the largest collection – some 740,000 samples – resides with the CGIAR's network of 15 facilities now operating under the ITPGRFA's multilateral system (CGIAR 2012).

Genebank collections vary significantly in condition, resources and composition, functioning within national and international arrangements.

---

8   Unless otherwise noted, data in this paragraph comes from FAO (2010b).

Canada's national genebank, for instance, has grown into a globally significant collection since its 1970 establishment in the wake of international encouragement. Plant Gene Resources of Canada (PGRC) currently operates as a network connecting two primary facilities – one for seed and one for clonal material – as well as smaller research centres focused on particular crops. In 1971 the centralised genebank held only 505 samples (Reid and Mosseler 1995), but PGRC is now one of nine national genebanks worldwide with over 100,000 accessions (FAO 2010b). Its collections include the world base collections for barley and oats, which together account for about 63 per cent of its total collection. In addition, wheat comprises 12 per cent, with the remainder of PGRC's collection made up of forages, crucifers, pearl millet, vegetables, natives, flax, clones and pulse crops (PGRC 2010). PGRC also serves as duplicate storage, particularly for US collections, and has deposited over 20,000 duplicates with the SGSV (NordGen 2012b).

As might be inferred from PGRC's or SGSV's collections, it is 'plant genetic resources for food and agriculture' that genebanking primarily targets. The ITPGRFA defines PGRFA as 'any genetic material of plant origin of actual or potential value for food and agriculture' (FAO 2009: 3). There is a distinction made in this definition between 'material' that contains heritable traits, and a more narrowly defined category of 'resources' in which material's traits have 'actual or potential value' for cropping. A range of PGRFA is stored in genebank collections, but a vast majority – an estimated 90 per cent (FAO 1997: 84) – is seed.

To become part of a genebank's collection, seed (or other propagating material) goes through a process of becoming an accession. This involves genebankers collecting, storing and documenting samples, after which collections must be maintained and may be distributed. When I visited the PGRC facility at the University of Saskatchewan for the first time, a poster on the wall outlined their process, which I have adapted and present in Figure 5.1. The worth of any genebank, given their dual purpose of conserving and making available for use PGRFA, lies in a combination of its material accessions and accompanying information, but also in sharing those resources. On these criteria, successive evaluations remark on progress made, but also on the continued need for improvement (*cf.* Brown et al. 1989; FAO 1997, 2010b; GCDT 2002). In the remainder of this section, a description of genebanking practice integrates with elaboration of some of its practical challenges – particularly regarding the partiality, loss and incomplete documentation of collections.[9]

---

9   Each genebank has its own procedures, constraints and capacities. This generalised overview draws on a well-developed practical literature including international standards (FAO/IPGRI 1994, 2011), as well as a 'best practices' handbook series, and online training modules developed by Biodiversity International. To this information is added data collected through genebank visits – both individual and group.

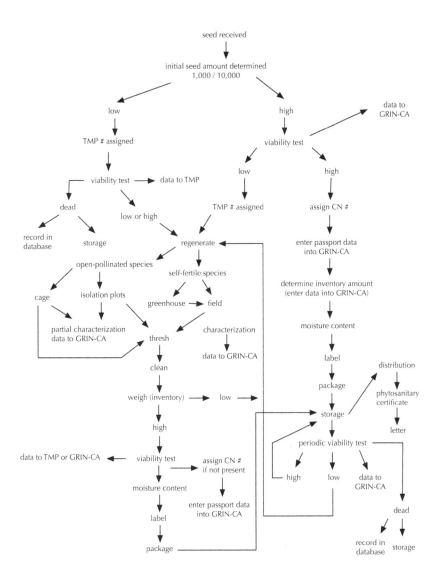

Fig. 5.1   Genebanking process at Plant Gene Resources of Canada
Adapted from: PGRC nd

Collecting

In genebanking, samples of seeds or other propagating material are either collected in-field – from farmers' fields, weedy edges or local markets – or acquired from other collections of plant breeders, genebanks or botanic gardens. Field collection depends on the target species and location, but as a rule collectors attempt to gather fruit or seed when mature and to garner representative samples of a population – something like 50 seeds per site (Engels and Visser 2003: 61). Ideally, initial collection makes multiplication before storage unnecessary, but often it is not possible to collect enough seed to meet this goal (FAO/IPGRI 2011). Field collectors encounter all the usual challenges of fieldwork – variable contexts and conditions, limited time and resources, incomplete information. Ensuring samples survive transport well can also be troubling, particularly with vegetative materials; over the years this has inspired new transport methods, from the Wardian case of the 1800s (a kind of portable mini-greenhouse) to the in-field *in vitro* methods of today.

The second option for acquiring materials – obtaining them from another collection – is the more common. Inter-institutional exchange is less costly and (usually) less time consuming, particularly if arrangements between institutions are already in place. Obtaining samples from other *ex situ* collections involves making a request, signing a material transfer agreement (a contract specifying terms of transfer, use, benefit-sharing, and often GE-free status), and waiting for delivery.

In either method of adding to collections genebankers must contend with multiple policies, and international acquisitions are particularly complicated. Biosafety protocols must be observed, ownership – involving state sovereignty and/or intellectual property rights – established, permission for collection obtained, and any benefit-sharing arrangements need to be negotiated. Some countries and genebanks have working bilateral arrangements, making this process easier. In addition, the Treaty's multilateral system and transfer agreement simplify this process, but only for specified crops and member countries.

Before collection of samples can occur though, genebankers identify, prioritise and target particular PGRFA. These decisions integrate genebank purposes and capacities with crop materialities, distributions and conservation status (see Barazani et al. 2007). For instance, a rare or threatened variety that the genebank can store and study properly will usually be prioritised over one already well conserved or for which the genebank has little capacity. These decisions rely on interest, resources and expertise at any particular time, to which are added considerations of previous documentations of species, sites and collections involving taxonomic classifications, collection characterisations, genomic analysis and mapping techniques. In short, genebanking requires complicated engagements even before seed samples are obtained.

The collections of PGRFA in genebanks vary by region and focus, but, in general, genebanks orient around conserving crops upon which humans and animals most depend for food (referred to as major crops). Cereals dominate global accessions, accounting for 45 per cent, of which over three quarters is wheat, rice, barley or maize (FAO 2010b). While neglected crops may seem incidental when viewed from a global food security perspective, many are regionally important for subsistence.[10] Despite some improvement in such neglected crops – yams for instance – overall, genebanking remains biased toward major crops.

In addition, accession types remain imbalanced. Known only for about half of total global accessions, accession types include landraces (44 per cent), breeding lines (22 per cent), advanced cultivars (17 per cent), and wild relatives (17 per cent) (FAO 2010b). By having so much invested in improved varieties (39 per cent), collections of other types become limited – despite higher diversity, numbers and need for conservation. Breeding lines and advanced cultivars tend to be more documented and ready for breeding (ignoring any IPR designations), while landraces and wild relatives often require additional assessment and development. Past breeding preferences have skewed collections toward improved varieties (FAO 1997; Fowler and Hodgkin 2004), but this appears to be shifting somewhat.

The last few decades demonstrate decreasing international field collections, understandable given combined challenges of in-field and legislative environments; however, over the last ten years or so domestic collections of landrace and wild relatives improved (FAO 2010b). These increased collections result from diverse factors. Previous international assessments and accompanying realisations that these seeds-plants are increasingly displaced have spurred interests in their conservation. Further, new methods make evaluation and integration of diverse materials faster and easier (Louwaars et al. 2009), which has fostered increased breeding interest that is enhanced by genebanks increasingly providing detailed information and molecularly-marked traits. Despite the genebanking of some of the most important food crops, a few examples of well-conserved wild relatives (including tomato), and recent efforts to fill collection gaps, however, much agricultural diversity remains in growers' fields (or their surrounding spaces). Accordingly, from the perspective of genebanking, much remains to be done.

---

10   The terms neglected, underutilised and marginal are often applied to varieties not globally significant in meeting food needs and that have, therefore, been *neglected* by researchers and genebankers, *underutilised* in commercial agriculture, and are, in this sense, *marginal*. Yams serve as exemplar, serving as a staple for an estimated 100 million people, but with only three breeders worldwide (Fowler 2004: 5; Mignouna et al. 2004) though varieties continue to be developed by growers (Scarelli et al. 2006). Yams are listed within the Treaty's included crops, and recently genebanking accessions have increased – from 11,500 in 1995 to 15,900 in 2008 (FAO 2010b).

Storing

Genebanks store seeds for longer terms by interfering with their metabolisms, which are slowed with dry and cool conditions. Before seeds can be stored in genebanks though, they are processed to ensure they meet parameters for quantity, accompanying information and quality – such as freedom from pests and diseases. If necessary, samples may first be multiplied, and then processed. In processing, the seeds are cleaned, tested, dried and then packaged for storage. How many seeds are stored depends on the species and expected seed use – between 3,000 and 12,000 seeds is recommended (Rao et al. 2006), with minimums of 1,500 and 3,000 for self- or out-crossing seeds in long-term storage (FAO/IPGRI 2011). Seeds may be packaged in special foil packets, as in the SGSV, but they are also put in cans, jars, boxes, paper envelopes and cloth bags. Whatever the container, it is sealed and labelled – often with a barcode – and stored.

Fig. 5.2   Storage at Plant Gene Resources of Canada

Typically genebanks store these accessions in differentiated conditions, reflecting genebank capacities and varied purposes. For instance, base collections – a set of accessions closely representative of original in-field collection – are often stored long-term for conservation. A subset collection of this base might be kept as an active, or working collection. Long-term storage, like that of the SGSV, will keep seeds up to an estimated 100 years and involves storing seeds at steady low temperatures and humidity – international standards stipulate -18 Celsius or cooler and 3–7 per cent moisture (FAO/IPGRI 1994). More commonly, seeds are stored in medium-term conditions (5–10 Celsius), which usually keeps high quality seed 20–30 years (FAO/IPGRI 2011). Freezers, fridges and cooling rooms are used for long and medium-term storage. Short-term storage, with ambient temperatures no higher than 25 Celsius, maintains seeds less than 10 years.

Each genebank's practices differ in relation to these standards; in Canada, long-term storage occurs at -20 Celsius or cryogenically, while medium-term storerooms are kept at 4 Celsius and 20 per cent humidity. Seed biology and quality affect longevity, as does the genebank's ability to maintain controlled conditions. Seeds may last only a few years if kept at room temperatures or in erratic states, but unreliable electricity, incomplete knowledge, malfunctioning equipment, limited resources and storage capacity, and events[11] all affect what can be accomplished.

If kept cool and dry, seed of most food crops – wheat, rice, beans, broccoli, among many others – will remain viable for years. But not all crops reproduce and/or store as easily. In rare cases, seeds will not survive the drying and/or cold conditions set by genebanking standards. Genebankers accommodate these 'recalcitrant' seeds – those of mango and cocoa for instance – by adjusting accession management by raising seed moisture, increasing humidity and temperature in storage, and/or regenerating seeds more frequently. Other crops – like potatoes, bananas and grape vines – reproduce best vegetatively. These crops are stored either as growing plants at field stations or *in vitro* – as plantlets in growing medium – in refrigerated storage. Some tissue samples may also be frozen in liquid nitrogen in a process called cryopreservation, which allows indefinite storage if material is properly prepared. In addition to seeds and plants, some genebanks also store (and may distribute) DNA samples or electronic genetic sequence information for research purposes (de Vincente and Andersson 2006).

---

11   As examples: Afghanistan's national collection was lost in 1992 to Mujahedeen fighters; Cambodia's diverse rice varieties disappeared with the Khmer Rouge's 1970s rural development schemes; Iraq's genebank was looted in 2003 sieges (Pearce 2005: 35); and in 2006 the national genebank in the Philippines lost 70 per cent of accessions to flooding and electrical shortages following typhoon Milenyo (GRAIN 2007).

To guard against possible loss, duplicate samples of accessions are taken and kept, sometimes in the same genebank but preferably in a separate location. While duplicate samples might become part of base or active collections at receiving genebanks, generally duplicates are 'black-boxed' – meaning the repository genebank stores the samples in long-term conditions, cannot use or distribute the seeds without permission, and is not responsible for monitoring or regeneration. Duplicates for long-term storage should contain enough seeds for at least three regeneration cycles, and the seeds should germinate at percentages higher than 85 (Rao et al. 2006: 94). Accessions sent to the SGSV are actually triplicate samples, serving as a back-up to this initial back-up, but the 'black box' stipulations are similar.

Duplication processes allow genebanks to mitigate possible losses; however, duplication brings its own problems. First, while duplicates can be accessed to restore collections, these samples are small and may be inadequate in times of rebuilding. Several generations of multiplication may be necessary to meet regular requests, not to mention dealing with any wider-spread seed needs in the wake of disaster. Indeed, this is (just) one of the critiques of the SGSV – that in catastrophe, for which it is preparation, seed would be too inaccessible and the amounts too small to provide for recovery (GRAIN 2008).

Duplication, in addition, results in redundancy worldwide. An estimated 1.8–2.2 million global accessions are unique, compared with 1–2 million in the mid-1990s, although total accessions have increased by 1.4 million (FAO 1997, 2010). This reveals that accumulations of similar accessions by genebanks exceed simple safety duplication. Viewed globally, duplication at these levels may seem wasteful. In fact, on this basis, the Global Plans of Action and the Trust advocate rationalising and streamlining genebanking collections (FAO 1996, 2012; GCDT 2004). But genebanks exist for diverse reasons, including research, breeding and even support of national food sovereignty. In fulfilling these mandates, internal duplication may be carefully managed allowing a genebank to do as much as possible, but globally unique accessions become less consequential. Genebanks, in other words, face conflicting objectives – their own and those of other interests – that prompt compromises for collections.

Documenting

Documenting in genebanks includes characterisation and evaluation practices. Characterisation begins with collection, recording information about what, when and where samples are collected. This might include, for example, the number of sampled plants, plant condition, habitat and particular locations – often with global positioning system coordinates. This 'passport data' is supplemented by descriptions of 'typical' seed and plant forms and traits – those consistently expressed by seeds that are easily observed and scored by assessors.

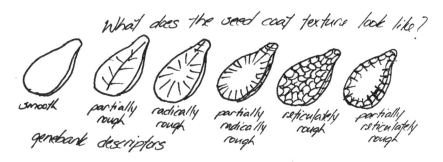

What does the seed coat texture look like?

smooth | partially rough | radically rough | partially radically rough | reticulately rough | partially reticulately rough

genebank descriptors

Fig. 5.3   Descriptors of seed coat texture

Source: Fieldnotes extract, based on Biodiversity International 2007

As part of projects supporting standardisation of documentation, Biodiversity International has collaborated with genebank curators and breeders to develop and offer guidance on producing descriptor lists for various species (see Biodiversity International 2007). Such descriptors include things like colour, height, texture and shape during different phases of life. A sample of seed coat texture descriptors can be found in Figure 5.3. Though some characterisation is done initially, further assessment can be done at any point while seeds are genebanked and may include molecular marking. Characterisation benefits potential users looking for particular traits, but also genebank managers' knowledge of collection composition and coverage, helping to identify needs for future collection and regeneration, or problems of over-duplication (Engels and Visser 2003).

Evaluation follows characterisation processes and involves examining traits considered to be of actual or potential use that vary with context – yield, stress resistance, and so forth. Some wheat evaluations, for instance, include assessments of 'bread quality' and 'resistance to lodging', each a useful trait for commercial applications (Reid and Mosseler 1995). As this example of wheat suggests, evaluation involves grow-out, and sometimes processing, of samples under particular conditions. The traits targeted for evaluation are often determined as part of public-private partnerships or client-based arrangements, and increasingly involve molecular marking. Results of characterisation and evaluation processes are recorded, sometimes in networked digital databases that are searchable online – such as Europe's EURISCO or Canada's GRIN-CA.

It is debatable whether genebanks should be responsible for providing information about specific traits of accessions given their limited resources, the reality that some traits alter with context, and that the benefits from such work often accrue to others. Indeed, some genebanks do not provide these services, releasing material samples with basic passport and (perhaps) characterisation data (FAO 2010b). However, genebank users desire detailed information about samples and their specific traits, and think that genebanks should provide it. In a study of the US National Plant Germplasm System, Smale and Day-Rubenstein (2002: 1649) report that despite generally positive responses from users to the data provided about the seeds they requested, the most common problem reported was 'inadequate or incomplete information about germplasm samples.' Further, the detailed information that accompanies improved varieties is one reason breeders rely upon them, despite acknowledgement of restricted genetic diversity and advantages of other types of seed (ten Kate and Laird 1999: 138).

Documentation and ensuring accessibility of information are widely accepted as necessary to promote collection use, but genebanks do not always succeed in these tasks. Much recent attention in training and programming promotes standardisation and digitisation for genebanking, and international efforts continue to connect existing genebank networks to a global system. And yet, a majority of countries do not maintain integrated national information, and several countries retain documentation – if assessment has been done – on hardcopy only (FAO 2010b).

Financial, language, skill and technology barriers play roles in restricting documenting accomplishments. Many genebanks, particularly national programs in the global South, do not have the resources to digitise information even if they have managed to document their collections (FAO 1997, 2010b; Smale and Day-Rubenstein 2002: 1640). In addition, things like common descriptor lists are not always available in relevant languages – a problem for several African countries in particular (FAO 1997: 121). Even among well-documented collections such as those of the USDA and CGIAR, collection comparison is inhibited by differences in naming systems and accession identifiers (Hammond 2011). Moreover, genebank and breeding objectives affect PGRFA documentation. As exemplars: wheat is well documented but not in similar ways; pulses and vegetables are poorly documented overall; and potatoes are only partially documented with select information available online (FAO 2010b). Overall, there are ongoing challenges with which and how accessions are documented, and in making that information accessible to users.

Regenerating

Genebanking does not stop after one cycle of collecting, storing and assessing material through which PGRFA samples become accessions; regeneration is required as seed viability falls or when stocks become low. While few remaining seeds makes need for regeneration or recollection obvious, knowing accession viability is more complicated.

As living organisms, even under optimal storage conditions seeds deteriorate, losing their capacity to reproduce. Periodic germination tests determine when accessions require regeneration. Genebankers must be careful not to over- or under-monitor accessions; the former wastes seeds while the latter misses declines in viability, and either results in unnecessary seed death. Guidelines recommend germination testing every 3–5 years using approximately 200 seeds, with caveats that species, seed quality, accession size and storage conditions each impact scheduling (FAO/IPGRI 2011). When testing indicates germination percentages below 85, seeds or plantlets are grown out, recollected and rejoin the genebank as new samples, which undergo an abbreviated process of becoming accessions.

Regeneration methods attempt to ensure purity and avoid genetic loss. Maintaining genetic purity might involve, for instance, isolating plants in greenhouses, distant fields or cages, while collecting from large populations helps safeguard wider genepools. The presence of GE crops – and their travelling transgenes – makes these methods both more difficult and more necessary when regenerating non-GE accessions. The standards endorse growing out at least 100 seeds, use of original samples and growing conditions as close to original collection habitats as possible (FAO/IPGRI 1994). Limited financial, infrastructural and staffing resources constrain regeneration activities, leaving genebankers with complex decisions requiring concessions (see Sackville-Hamilton and Chorlton 1997). For instance, genebankers sometimes combine characterisation, evaluation and regeneration processes. However, this is not ideal since regeneration requires different growing conditions to ensure genetic integrity, health and viability compared to those for specific trait evaluation (Engels and Visser 2003: 72).

Genebanks cede accessions to natural and political events – typhoons and wars for example – but most seeds fall to a more banal agent: neglect. Genebanks around the world admit chronic regeneration failures, but report that they are simply unable to meet accessions' needs due to lack of funding, infrastructure and trained personnel (FAO 1997, 2010b).[12] To alleviate regeneration pressures some genebanks, including Canada's, enlist breeders, private seed companies and non-governmental organisations in regeneration efforts (FAO 1997). In one such program, Seeds of Diversity Canada's members regrow seeds, sending seeds and documented results back to PGRC, while keeping some seed for themselves in exchange. There are benefits to these arrangements for genebanks, but they bring their own challenges – consistency of condition and information in

---

12   In 1996 of 95 countries, 71 reported regeneration problems – affecting approximately 60 per cent of worldwide accessions (FAO 1997: 112). The average regeneration need was 48 per cent. Since 1996, regeneration capacity decreased for 20 per cent of genebanks and increased for 18 per cent, with the rest remaining at similar levels (FAO 2010b).

particular (Diederichsen 2012; Richards 2007). Further, despite such measures, regeneration backlogs continue to plague genebanks.

As part of its international mandate, the GCDT supports regeneration programs for targeted collections in global South genebanks, getting samples in shape for storage in SGSV. But sometimes it is too late; in addition to the 74,000 'rescued' accessions, the Trust reports finding 12,000 already dead (GCDT 2011). This is neither a new problem, nor one unique to global South genebanks. For example, soon after establishing the national genebank the Canadian government voiced concerns about maintaining viable accessions, which raised thoughts of appealing to IBPGR for aid (an option rejected as a national embarrassment) (Fowler and Mooney 1990: 166–7). Ongoing difficulties of maintaining collections led to the decision in the 1990s to adopt PGRC's multi-nodal structure; however, many accessions had already been lost (Richards 2007). This kind of die-off is one of the reasons that some people, arguing instead for in-field saving, refer to genebanks as seed morgues or cemeteries (Ausubel 1994: 82). This is somewhat unfair considering the continuing losses in-field, and seems harsh given the best intentions and efforts of genebankers despite limited resources. However, the loss of collections due to neglect – unintended though it may be – remains significant.

Distributing

Most genebanks, unlike the SGSV, distribute samples of their accessions. In general, samples are requested and sent out to breeders and scientists – public and private – researching and developing crop varieties. Distribution among genebanks is also common; these exchanges may be repatriations mandated by international or bilateral agreements, but most often they are part of regular acquisition. In rare instances, genebanks may supply growers directly with seed for cropping (Fowler 2008), but usual practice offers indirect benefits to growers in the form of new varieties bred by those requesting genebank materials. A number of factors interfere with distribution: the size and quality of accessions; budgets; information available to potential users; and policy environments.

Few genebanks publish details on who accesses collections and why. However, the CGIAR reports that of the average 100,000 samples their network distributes each year, most exchanges occur among their research centres (48 per cent), followed by delivery to developing countries (30 per cent), developed countries (15 per cent), and the private sector (3 per cent) (FAO 2010b). This data is helpful, but limited. Consider, for instance, private sector access. First, this sector possesses its own collections – built on previous years of accessing genebanked materials, breeding varieties and purchasing other private collections, which mitigates the need to access current public collections (ten Kate and Laird 1999: 137–42). In addition, private capital participates in public-private partnerships with genebanks, universities and government breeding programs. It is these public institutions that are likely requesting initial research materials, therefore corporations would not be noted in the data despite their involvement. Moreover, the CGIAR operates within the Treaty's multilateral system, which skews distribution patterns. Corporations, given the option, source PGRFA accessions without the Treaty's IPRs prohibitions – relying instead for example on the US National Plant Germplasm System (Hammond 2011). With limited information on the flow of materials – from initial requestors and beyond – the particular interests served by genebanks remain uncertain.

To sum up, genebanking works at accomplishing conservation and availability of plant genetic resources – especially those most useful for human and animal foods. Collecting, storing and documenting processes transform seeds into genebank accessions, while regenerating and distributing reconstitute and spread accessions. In achieving their dual mandates, however, genebanks face challenges. In practice, genebanks conserve select materials and are not always able to meet international standards, seed needs or user demands. Though always evolving, these challenges are not new or even newly recognised; rather, the 'predicted pitfalls' of biased collections, inadequate management and restrictive political environments have been endemic and analysed since the beginnings of genebanks (Pistorius 1997: 23). Accepting for the moment genebanking objectives, much of the limitations covered in this section might be dealt with by increasing and targeting funding, improving infrastructure, employing adequate personnel with appropriate skills and establishing clear (and facilitative) policies. Moreover, advocates of the SGSV suggest that the Vault does much to address genebanking challenges, in combination with the wider programs of the Trust and the Treaty (Fowler 2008). However, other criticisms of genebanking reveal more fundamental constraints and implications, and it is to some of these that I now turn.

Questioning Genebanking

Genebanking practices currently shape and are shaped by other seed-people relations and realities, implicating genebanking in efforts of seed reordering. In considering how the technoscientific practices of genebanking matter, science and technology studies offer inspiration. Science studies scholars have illustrated the passionate practices of scientists (see Fox Keller 1985; Haraway 1989; Latour and Woolgar 1979), and genebankers display no less sense of attachment or obligation. But respecting this does not exempt science or its methods from scrutiny; its import and entanglements in contemporary ordering demand attention.

In her inquiries Stengers (2000, 2010) offers cautious celebration of scientific accomplishment – lauding creativity and experimenting, while insisting on examination of particular practices, obligations, affects and futures made possible (or excluded). Science, and the knowledge it generates, is, in her view, accomplished through and generative of particular practices. In this way, universalism is denied – to science and to those who study its practicing. Genetic engineering serves as one example for Stengers's argument that the entanglement of science with neoliberalisation and corporatisation results in impoverished techno-science. Indeed, on this basis, she excludes biotechnology from consideration of scientific practice. This concern with science in the service of private capital involves a dual reduction: first, impoverishing inquiry by shifting focus from experiments to commercial outcomes; and, with this reorientation, the phenomena being studied is reduced to consideration of rules objects follow. Stengers' insights might be productively related back to arguments made in Chapter 3 about genetic engineering, but here I wish to extend her ideas to genebanking to think about the knowledge created and shared, the objects included and enacted, and the practice's relations.

In this section, keeping Stengers in mind, I consider three underlying ethico-political issues with contemporary practices of genebanking. The first involves genebanking's entanglements with existing inequities among humans, particularly in relation to reinforcing divisions between farmers and breeders and between global North and South countries. These divisions reflect a privileging of scientific expertise and commercial development, resulting in inequitable access and benefit-sharing in relation to PGRFA. The second issue relates to the reconstitution of seeds through genebanking, such that these complex nonhumans become natural resources to be conserved in stasis, serving private capital priorities. The final concern regards the possible perpetuation of genetic reductionism through genebanking.

These issues exist in partial connection with those outlined in earlier chapters, working with other practices in re-establishing divisions between nature and culture as well as revaluing living organisms and their relations in ways advantageous to commercial interests. But genebanking differs in its particularities of how it affects and is affected by seeds, savers and saving. In addition, each of these critiques of genebanking presents to varying degrees, and it is worth remembering that genebanking – like genetic engineering or encoding seed-related rights – might be practiced and related differently. Some efforts at fostering alternative genebanking – more and less successful – are noted throughout the section.

Reinforcing Divisions and Inequities

First, is the issue of genebanking's ties to and reproduction of existing inequities within seed-saver ordering, which presents as divisions between breeders and farmers as well as between global North and South. The first of these involves the close relations exhibited between genebanking and (increasingly private) breeding. The conservation of crop diversity is not promoted only for conservation's sake, but also to ensure that seeds are available for use. The shift beginning in the early 1900s to laboratory breeding of pure lines established assumptions about what useful material and information meant and to whom access should be granted. In this context, genebanks evolved, at least in part, to serve commercial breeding cultures and that influence carries through in contemporary practice. Genebanking assists commercial breeding better than seed saving in several ways: first, genebanked seeds are kept (as much as possible) in stasis, preserving specific traits, while seeds in-field continually alter; second, genebanking centralises seeds in relatively few locations making material easy to access, while seeds in-field are widely dispersed making collection more expensive and difficult; and third, genebanking includes collecting and distributing (at least) basic scientific information along with the seeds in question, while with *in situ* seeds information may be limited to ethnographic accounts of local populations.

In contrast to genebanks serving breeders (and other researchers), growers are presumed to benefit from genebanking indirectly, through the commercial breeding of new varieties, rather than, as in earlier collection programs, through direct receipt of seeds for in-field experimentation and development. This assumption is reflected not only in genebank distribution policies restricting access to 'legitimate users' (i.e., breeders and researchers), but also in international agreements. For instance, the Treaty privileges scientific expertise and access by defining 'access' as availability for research, instead of offering a broader meaning. As suggested in Chapter 3, however, growers' needs are not necessarily met by breeding within the corporatising seed order, and its orientation toward worldwide use and commercial profitability. Moreover, in-field saver-seed relations are responsible for most crop diversity. Savers are estimated to breed and adapt over a million varieties each year, compared with the 10,000 or so varieties with PBR protections developed by breeders (Dutfield 2011). Indeed, as one USDA researcher noted, 'soybean breeders must admit that a more ancient society made the big accomplishment in soybean breeding and that we have merely fine-tuned the system to date' (as in Kloppenburg 2004: 185). Moreover, based on a 2004 inventory, two thirds of the varieties grown by saver members of Seeds of Diversity Canada were not available commercially or conserved within genebanks (Wildfong 2005). Yet much genebanking excludes savers, who remain undervalued as contributors, potential users and experts in their own right.

In some ways, PGRFA governance and genebanking are paying more attention to the importance of saver-seed relations. For instance, countries are urged to participate in supporting farmers' rights through the Treaty, and the Special Rapporteur on the Right to Food calls for support of saver participation and networks as part of sustainable agrifood orders (DeShutter 2009). Further, some genebanks have special projects working with growers to achieve more meaningful results – including participatory plant breeding (Engels 2001; Feyissa 2006b; Labrada 2006; Morris and Bellon 2004; Vernooy 2003; Worede et al. 2000). Cooperative programs with savers – such as the regeneration or storage initiatives mentioned earlier – also suggest potential for genebanking to become more grower-oriented.

Possibilities, therefore, exist for expertise to be more distributed such that it is not only scientific experts who are allowed to know, speak about or practice breeding or 'banking'. Different ways of knowing seed worlds may be accepted instead of privilege being accorded automatically to one. However, these initiatives remain underdeveloped and inadequately linked with genebanking. Overall, standard genebanking, and the expert knowledges it produces and supports, retain higher levels of valuation and attention. The calls more easily heard in international and many domestic arenas are those endorsing closer ties between genebanks and their existing users, as well as for increasing public-private partnerships in research and innovation. For instance, both the FAO (2010) and the Trust (2011, 2012b) advise genebanks to form closer user relations and expand historical services to breeders by engaging in molecular marking and pre-breeding, each of which serves commercial interests in having PGRFA easily accessible and quickly finished for market with limited expense. These calls, added to private-public partnerships and corporate contribution to initiatives such as the Vault, build upon historic relations and tighten genebank entanglements with the corporate seed order.

In combination with reinforcing boundaries between savers and breeders, the genebanking system maintains historic social inequities between global North and South countries. Most agricultural diversity was developed by farmers in the global South, and this diversity forms the base of genebank collections and breeding programs – public or private. Yet when seed samples were collected and stored, growers' knowledge and seed relations often went unacknowledged. This is part of what Plumwood (2001) calls a 'profound forgetting' of complex histories; it is a forgetting that isolates seeds from their relations in the service of particular interests. Often this collected crop diversity has been further developed, or simply adapted to new growing conditions, proving economically profitable for breeders, nations, and/or private interests. Legal options to claim IPR serve to complicate these already inequitable relations, reinforcing not only profit-making possibilities but also lack of trust among countries and between farmers and seed breeders or bankers.

Whether referred to as biopiracy (Mgbeoji 2006; Shiva 1997) or bioprospecting (Scholz 2004), these exploitative practices go far back in history and impacts can be significant. In economic terms alone, it is estimated that just 2 per cent of royalties for pharmaceutical contributions of varieties would amount to USD5 billion for the global South, with another 300 million garnered from agricultural products (Carolan 2010: 28).[13] The CBD and the Treaty include attempts to recognise savers, countries of origin, benefit-sharing, and to limit IPR. But aside from the caveats and limited implementation that attend each measure, several already mentioned, these provisions do not apply to materials extant in collections or developed before agreement enactment – by which time much had been accomplished.

Though representatives sometimes participate in discussions, genebanks do not set international or domestic policy agendas – public or private. The material and discursive enrolment of genebanks in contemporary seed reordering, however, facilitates and reinforces inequities between farmers and breeders, and between 'gene-rich' and 'technology-rich' countries. These divisions reflect a particular valuing of scientific expertise as the best means to achieve food security through improved seeds, and a commercial interest in profiting from agrifood. Efforts in international and national forums – particularly by non-governmental organisations and states promoting food sovereignty principles – continue to push further in addressing these more human-centred inequities.

Reconfiguring Seeds

The second underlying concern with genebanking relates to the ways in which the practice reconstitutes seeds – turning them into PGRFA stored ready-in-wait for technoscientific development. In the first instance, genebanking concentrates upon material that has the capacity to reproduce. In addition, that material must display heritable traits, and these traits must be useful (or potentially useful), particularly for consumption. Genebanks, in other words, work with plant genetic resources – mostly for food and agriculture. This is part of the legacy of close ties to breeding agendas, which today carry with them neoliberalisation and private capital priorities. By defining what is to be genebanked in such a manner, inclusions and exclusions are made real. Like the object definitions made through IPR (see Chapter 4), seeds (and other plant parts) are conserved if they meet the criteria, if they do not then other resources

---

13   In comparison to pharmaceuticals, food crops tend to have multiple parent lines (making country of origin difficult to determine and delineate), to have shorter (and often less profitable) commercial lives, and to come from materials within private collections and/or acquired before CBD or ITPGRFA enforcement (Andersen 2008).

are sought. Moreover, the information created and distributed by genebanks about PGRFA orients around useful traits, increasingly indicated by molecular marking. This reconstitution of seeds reflects and supports commercial remaking of seeds into commodified delivery mechanisms of useful traits.

In order to be genebanked, seeds are removed from their context, a move deemed necessary for their conservation and sustainable use. In addition to isolating seeds, cryopreservation methods and the SGSV's goal of conserving 'in perpetuity' suggest how genebanking works toward an ideal of static PGRFA. A division between nature and society is reasserted here such that seeds (belonging to a nature 'out there') can reasonably and feasibly exist as useful static entities – seeds become PGR. In this context, Van Dooren (2009) proposes that the PGR of genebanks serve as proxies for more complex, embedded, in-field seeds, arguing further that these 'good enough' versions result in impoverished conservation. This idea of seeds genebanked as proxies helps to account not only for limits in accomplishing biodiversity conservation, but also for efforts to eliminate duplication (since several of the same proxy is unnecessary) and limitations in meeting growers' needs. The transformation of seeds within genebanking, however, is more than a move from complex beings into PGR(FA); PGRFA is, in its turn, converted through genebanking into accessions, and it is accessions that are most useful to breeders. For breeders, genebanked accessions are better than, rather than poorer representations of, those complex organisms in fields – accessions are centralised, convenient, clean, already isolated, and come with useful information. The same set of practices, therefore, exists differently depending on its encounters, and only for some does genebanked seed stand-in for better seed.

The vitalities of seed challenge genebanking practices – as suggested with chronic regeneration failures. Genetic sequence information remains insufficient as PGRFA – as it does not allow reproduction – and therefore, living materials, with all their capacities and dispositions, continue to interfere in genebanking. There is some concern, for instance, that just as seeds adapt to other places, that they may adjust to living in genebanks – getting used to long periods of dormancy, altering genetic profiles or preferring particular growing conditions (like greenhouses) (del Rio and Bamberg 2003; Diedrichsen 2012; Gómez et al. 2005; Richards 2007). Methods to keep PGRFA as close as possible to its condition at original field collection, to slow the need for regenerations, and to catalogue particular traits all indicate the drive to constrain seeds' perceived excesses in order to allow for accession futures. Static PGRFA remains (so far at least) unachievable, and may not even be desirable.

In comparison to genebanking, seeds saved in-field continue as part of changing socionatures, altering with diseases, insects, wild relatives and climate as well as growers' practices, cultures and knowledges. In early discussions of genebanking, Bennett (1968: 63) warned that distancing genebanks from saver-seed relations would be counterproductive to conservation and sustainable use, explaining that *in situ* collections accomplish these goals otherwise:

> I think that attempts to find other merits in the 'steady state' which seed storage represents, seem to come dangerously near to museum concepts. The purpose of conservation is not to capture the present moment of evolutionary time, in which there is no special virtue, but to conserve material so that it will continue to evolve. Such 'continued evolution' could only be possible in *in situ* collections.

Bennett's early articulated concerns still echo in advocacy of understanding seed-saver changing relations in positive ways, in ways complementary to genebanking and deserving of serious attention in their own right – for instance through farmers' rights or in-field conservation.

There are some efforts to acknowledge and respond to agricultural diversity in more complex ways and to include in-field conservation. Both the CBD and the Treaty include *in situ* conservation as an obligation for states, and the Global Plans of Action for PGRFA recommend stronger connections between *in situ* and *ex situ* efforts (FAO 1996, 2012). Further, the FAO's (1998: Sec2.2) redefinition of agrodiversity, for instance, recognises previous thinking about crop diversity was too simple, modifying the understanding to include genetic, species and ecosystem diversities that vary with environmental interactions and agricultural practices (see also Crucible II 2000). With this broadening of definition, boundaries between nature and culture become blurrier, while in-field conservation and connecting with savers become both appealing and necessary. Closer, more complementary, relations between genebanking and saving may be desirable, but so far this remains a relatively minor part of genebanking, as well as broader international and national conservation and sustainable use initiatives. Taking better advantage of these opportunities is an ongoing task.

Considering the Gene in Genebanking

Each of the two previous concerns leads to a third consideration: how seeds are learned and known through genebanking. Practicing genebanking entails learning to understand and respond to seeds through established technoscientific 'best practices', which – as the name implies – involves attending genes. Scientific knowledge of crop diversity involves various levels (genetic, species, ecosystem); however, it is with genetics that diversity of seeds (and other living beings) is understood to ultimately lie (Takacs 1996). As noted, in the early 1900s commercial interests and rediscovery of Mendel's laws facilitated a breeding shift from adapting varieties in fields to adding desired traits through laboratory breeding. 'Pure line' methods developed insights about 'factors of inheritance' (i.e., genes or genetic sequences), moving breeding into the natural sciences realm (as opposed to agricultural practice) (Pistorius and van Wijk 1999; Kuyek 2004b). Building on ideas of seeds as separate natures serving as resources for humans, the interest in traits and inheritance added to seed refigurations.

Even early discussions of long-term storage and conservation were immersed in scientific method and genetic resources. For instance, the handbook that emerged from the FAO's 1967 meeting states that 'efficient utilization of plant genetic resources requires that they are adequately classified and evaluated' (as in Pistorius 1997: 21). As in-field diversity declined with adoption of 'improved' varieties, genebanks became important mechanisms to ensure genetic resources were conserved and made available for breeding purposes. The prioritisation of genebanking not only facilitated particular breeding practices, but provided a sense of security; even as diversity disappeared in fields, future food security was assured. In the process of favouring genebanking over seed saving in fields, those actors with large collections gained more influence, as did breeders. Scientific lab-based breeding and genebanking of PGRFA – reproducible material with useful inheritable traits – became the supported norm.

Critics suggest that the power of determination attributed to genes may be appealing in its simplicity and clarity, but offers inadequate understanding (Fox Keller 2000; Oyama 2000). McAfee (2003) describes this overvaluation of genetics as molecular-genetic reductionism, in which genes are constructed as discrete entities that cause (alone or in concert with other genes) particular traits to manifest in organisms. Genes, in this view, decide the rules seeds obey. In the application of this approach through genetic engineering for instance, genes, once properly understood, can be studied, altered, transferred and turned on or off to serve particular purposes. In a sense, this allows not only the rationalisation of natural worlds, but ensures nature is 'made to order' (Bamford 2002: 46). Further, as discrete entities, isolated genes or genetic sequences, combined with the technoscientific processes used to study and alter them, become potential private property and profit-makers. Again, seeds are sliced into, divided, reformed, and/or lost in the competition for marketable traits. By knowing seeds through genetics and through enacting the conserving, assessing and making available of PGRFA, genebanking helps make this reality possible. While genes and traits do occupy much of genebanking, it is somewhat unfair to suggest that seeds are lost and genes understood as deterministic – hidden and foundational might be more reasonable descriptors. When I asked the curator of PGRC what he thought of the genetic reductionism criticism in relation to genebanking, he responded by suggesting that 'genebank' as a term emerged within a particular historical context, predating much concern with genetic engineering and genetic determinism, and perhaps this context was being missed. Further, he explained that although genes matter, 'the base unit for us as a genebank is the seed' (Diedrichsen 2012); understanding genetic diversity with molecular science was just one approach that combined with agronomical and morphological methods. Moreover, although he did not connect previous statements supporting collaborations with savers, his recognition of ongoing contributions of savers to conserving and sustainably using seeds was clear, suggesting a more complicated view of seeds and their relations than perhaps allowed by some critiques of genetic focus. Genebanking, to this genebanker and several others with whom I have spoken, is about more than the genetics of PGRFA accessions.

In considering genebanking and a bias toward genetic diversity and PGRFA, I do not suggest that genes do not participate in organisms' biologies or that scientific knowledge does not contribute meaningfully to understanding seeds and their relations. Rather, the aim is to refuse universal or unitary understandings. It is, in other words, necessary to acknowledge that a complex mix of actors and processes – only some of which are genes or gene-related – account for any one organism's living (Fox Keller 2000) or genebanking practices. Moreover, it needs to be recognised that conserving and making available PGRFA makes ethico-political differences. Genetic diversity understandings and genebanking practices are useful. Indeed, given rates of extinction, humanity's survival may depend upon the seeds and knowledge collected in genebanks; certainly, much more crop diversity would already have been lost without genebanking. But genetic approaches, genebanking, and science more broadly, account for only part of lived realities and agrifood futures.

Genebanking involves scientific experiments, particularly in evaluation processes, but much of the day-to-day operations do not risk knowledge in the way that Stengers suggests for science. Rather, standards are followed using established techniques and equipment, the questions posed are not always creative, and research often reflects commercial and governmental priorities. In this way, genebanking might be considered 'impoverished'. But Stengers (2000: 98) argues that part of what she hopes for is a shift in question from 'which variables do you, as an object [like seed], obey?' to 'which variables will you take into account?' The former question sits easily with a focus on genetic traits that can be marked, characterised and modified, and that make seeds obey. But the latter question requires researchers to allow those studied to interfere in experimenting results – it accepts some seed agency and demands accommodation of their vitalities and relations. This, along with her conviction that agentic nonhumans are those who/which make us think differently (Stengers 2010), seems to be reflected in genebanking practices – at least some of the time. As the PGRC curator reflects, 'it is living material that we preserve in genebanks, it is not dead material. And I think it is fair to say that living material will always be in change' (Diedrichsen 2012). Seed vitalities provide fascination and challenge for genebankers. The problem of keeping seeds alive, but stored for long periods, makes genebankers think about what is best for seeds and how they will respond to their genebank lives and beyond them. Moreover, as the section on practicing illustrates, finding, assessing, storing and regenerating seeds means paying at least minimal attention to their capacities and dispositions – as whole organisms as well as carriers of traits.

Summing Up

By the time the term biodiversity was coined for the 1980 *National Forum on BioDiversity* in the US, the international community had already recognised the need for conservation of crop diversity in genebanks and established organisations and methods to study and oversee such efforts. Genebanking did not emerge in isolation; rather, it was (and is) part of complex relations and orders linking seeds, ecosystems, knowledges, institutions, and various human interests. The evolution of genebanks' dual mandate to conserve and to make available useful material points to this complexity.

The practical challenges of genebanking well combine with more fundamental ethico-political questions raised by the practice and its relations with other seed-people worlds. A sense that genebanks provide security enables 'business-as-usual'; there is less need to alter our ways of doing agrifood (or even question them) if back-ups and scientific expertise are at-the-ready. However, genebanking's practical and ethico-political issues should unsettle this sense.

Disruptions of genebanking achievements also remind us of the importance of seed saving practices and present possibilities that genebanking be done otherwise. Genebanks have accomplished much, and we may well continue to rely on them for seed and information. Rethinking and reconstituting them as complementary to saving and operating within broader programs working toward more sustainable agri-cultures opens different options and challenges. This chapter is the last of three revealing seed-people arrangements with neoliberalisation and corporations, regulations and intellectual property rights, and genebanking processes. In the next three chapters, the shared worlds of seeds and savers provide the focus.

Chapter 6
Learning Seed Saving

I spent yesterday learning how to save seeds on a farm about two hours from Toronto. It was overcast in the morning and I thought it might rain, but luckily it held off until early evening. Guided by [an experienced seed saver, name omitted], I helped gather and clean several different seeds, plant some garlic and bury some biennials for an experiment in overwintering. My first task was to harvest marigold seed from several rows planted as an arch running up and across a sloped field. The marigold flowers were almost all gone, with a few late bloomers hanging on. Saving these seeds was pretty straightforward. I was told and shown how to separate the seeds, and given a brown paper bag (labelled) to put them in. I experimented with the best way: do I sit, bend or squat?; which hand does what?; which flowers have lost their petals and are dry and open enough to pour seeds out, which will need to be torn?; what's the best motion to separate the petals if they're still attached?; which are likely to have fully formed seeds, which will be passable, which not worth picking? After a while I figured out what worked for me and got a rhythm going: pluck, tear, pour, drop; repeat; shift position as necessary. It was quite satisfying to see the bag slowly filling with seeds for next year's planting. I did that for a while, and then we did other seed saving. I really enjoyed it. I look forward to doing it again. ... Today sore muscles in unexpected spots remind me that I do not often spend my day that way. It made me laugh at myself. (journal extract)

This retelling of my first saving of marigold seeds highlights seed saving as a set of practices that involves skill developed through experience with seeds and ideas from other savers. Through saving seed I learned to sense myself and seeds differently. The word 'sense' here acknowledges two different, interrelated meanings: first, understandings; and second, feelings and particular physical senses – touch, smell, hearing, sight and taste. In this chapter I argue that through practice with seeds, seed savers learn to expand their 'sense-abilities' – their abilities to sense and make sense – and that through these experiences attachments form.

In the discussion that follows, the first section deals with conceptions of how learning opens (and closes) possibilities. This includes an elaborated example of learning with carrots, signalling how saving offers and requires different – but related – engagements in comparison to eating and growing. After this first section, I focus on experiences of savers in sensing and valuing senses; living with seeds and learning their rhythms; feeling a sense of wonder and attachment; becoming skilful practitioners through experiment; and engaging as part of human and nonhuman collectives. Throughout, learning saving is explored as a material, sensory, cognitive practice in which humans and nonhumans collaborate in learning and saving.

Learning Practices

The idea that learning or training opens up possibilities has lately been elaborated in concepts of 'learning to be affected' (Latour 2004b) and becoming 'response-able' (Haraway 2008). Latour argues that through training a person becomes more able to attune to and understand the many differentiations of our worlds. As an example, he describes the process of becoming 'a nose' through relations of a student, teacher and odour kit. In his example, it is through training that the pupil gains a nose, sensing differentiated aromas previously undetectable and, in the process, learning to 'inhabit a richly odiferous world' (Latour 2004b: 225). Latour insists that this process of learning to be affected is not unidirectional or isolated; rather, bodies, languages, technologies co-constitute a dynamic and variegated world.

This conception of learning to be affected is useful in thinking through seed saving by: beginning with bodies and senses, making it clear that learning and knowing are embodied and situated; interrupting a strict dualism between subject and object (human and odour kit) that would interpret seeds as static objects upon which humans act; and focusing on process and practices, highlighting dynamism and possibilities of change through experience and experiment. But somehow Latour's learning to be affected feels too distanced; despite the embodiment in the encounter, some of the sense of the experience is missing.

Haraway's (2008) exploration of companion species illustrates learning relations-in-the-making in ways more personal, ethical and political. About her and her dog Cayenne's relationships in training to play agility sports she says, 'both players make each other up in the flesh. Their principal task is to learn to be in the same game, to learn to see each other, to move as someone new of whom neither can be alone ...' (2008: 176). Here both players – dog and human – train together, learning and forming attachments. Opportunities appear and disappear in training as participants change each other and their shared practices. Haraway's accounting of learning-in-relation is clearer about lost opportunities, fleshy living and ethico-political implications, but both Latour and Haraway agree that experiential learning involves cultivating and changing senses, skills and attachments.

As humans and nonhumans come together in practice then, they affect and are affected by one another, and their worlds. Further, it is through these shared and expanding experiences that a sense of attachment to others and shared worlds forms, affording opportunities for 'receptivity and generosity towards other bodies' (Bennett 2001: 158) and reconstituting a 'feeling life' – a grasping of life combined with attunement to it (Thrift 2000: 46). It is this kind of approach to learning among seeds and savers that finds explication in the following pages, but first I offer, by way of example, some reflections on a particular food-plant-seed.

Knowing Carrots

Let's start with carrots. 'A day is coming when a single carrot, freshly observed, will set off a revolution', said Paul Cézanne. He would not have been speaking of saving, growing or eating, but of art – specifically of the post-impressionist approach shifting art worlds. But we can still take a lesson: paying serious attention to seemingly insignificant things, sensing them anew, alters worlds – maybe even in revolutionary ways. As Law and Urry (2004: 391) suggest, 'to change our understanding *is* to change the world, in small and sometimes major ways.' Painting for Cézanne was lived experience rather than objective representation, and he aimed to convey a sense of that living through his painting. But let's get back to carrots – and the learning of them.

Learning and knowing carrots might be related first to eating practices. Carrots and consumers' knowing of them provide material for Roe's (2006a) investigation of edibility. Roe's study, though primarily concerned with how GE foods become inedible, includes analysis of group discussions about edibility of three kinds of carrots: tinned; young and bunched; and, fat, mature. Taste is individual, she finds, but carrot materiality is important overall in determining edibility. How carrots look and feel – their appearance and textures – are clearly sensual and easily related with consumers' material experiences, but 'invisible' properties – like vitamin levels or quality of sunshine – are also inferred. For her participants, how orange or how old a carrot appears, for instance, implies the carrot's production processes and nutrition. Roe further argues that consumer considerations are both haptic and cognitive, integrating senses with memories, practices with preferences, and so on. Consumers judge whether they will eat carrots, and how, by material, affective and cognitive engagements (see also Eden et al.'s (2008) 'mucky carrots').

Carrots are not only those things we're used to eating though, they have other experiences and other ways of knowing them. As Roe (2006a: 474) notes, '[c]arrot bodies, even when diced, sliced, boiled, fried, have a "sameness" that stands out more than that process by which it developed from seed to carrot shape.' If we want to know some of this carrot difference, we need to look at other carrot–people practices such as growing.

Second, then, learning carrots may manifest in growing practices. Pollan (1991: 118–20) recounts his own quest to grow carrots well – something he struggled to do season after season. In trying to understand what is going wrong with his carrots, asks himself, 'what does a carrot care about?' and considers 'what matters to them, what they require in order to fulfil the terms of their destiny.' After much careful thought and through experimenting with varied growing practices and spaces, the problem is found in heavy soil and crowded carrots. Pollan then goes about providing his carrots 'a propitious place by lightening the soil', 'ruthlessly' thinning the seedlings, and finally growing carrots of which he is proud. Through his experience of learning carrots, Pollan develops skills that will now allow him to 'grow fine carrots without a moment's reflection.'

Roe's eaters and Pollan's growing reveal how practicing with carrots involves learning to be affected through experimenting with carrot living, illustrating how learning is part of the bodies, spaces and times in which it develops. Practical outcomes – carrots to eat – are only part of the point; through these practices, other implications take shape. First, as Roe suggests, eating – and one might add growing and saving – any carrot is ethico-political because the carrot 'embodies material connections that leads one back to the workers in the carrot factory, the field where the carrots were grown, and the environmental changes to flora and fauna that produce a carrot mono-culture' (Roe 2006a: 478). Roe here makes an assumption that the carrot is an industrially produced one, but even a carrot grown otherwise carries with it ethical and political implications, making differences. Moreover, carrots influence these relations by virtue of their histories and materialities. Both Roe and Pollan are cautious about anthropomorphism, but carrots are allowed some agency through acknowledgement of their capacities, tendencies and embodiments. Growing and eating good carrots, then, is collaborative and ethico-political, remaking carrots and people in particular ways. People learn carrots through eating them, growing them and more.

Seed saving has its own carrots. The challenges faced in growing carrots well are just part of those of seed saving, as savers attempt to learn the carrot-ness and the seed-ness of carrots. Appearance, integrity and invisible materialities – as outlined by Roe (2006a) – are also involved in seed saving. When saving for food carrots, you choose varieties that are good to eat. And once you grow them, you make sure not to eat them all, because you want seed from some of those carrots. Or, if saving for conservation, you might select based on 'typical' variety descriptions, doing your best to conserve genetics unknowable by material senses alone. In the case of carrots, seed saving offers not only a more drawn out and involved relation, but also opportunity to experience carrots in their growing-to-seed and seed forms.

Carrots are a crop that some savers shy away from because they are biennial – producing seed in their second year. A few savers I spoke with shared stories of being surprised by carrots growing the next year when harvest was incomplete, prompting saving of carrots that would not have happened otherwise. In general though, carrot saving obliges investment. Some savers use simpler methods – leaving carrots to overwinter and saving seed from those that endure – but savers concerned about having good tasting and growing carrots, producing quality seed, and/or biodiversity conservation usually take further steps.

Dealing with carrots in their first year is similar to growing other seed crops: monitor, mark those for seed, harvest, select those best suited for saving, eat those remaining. But then the saving process continues. Seed carrots demand trimming, over-winter storing, cold-treatment shocking, retrimming and reselecting in spring. Those carrots that make it through are replanted and grow specifically for seed saving – no food carrots this time around!

Fig. 6.1   Carrots going to seed in their second year

In the carrots' second season, their habits are tracked again. And, since they have plentiful cultivated and wild relatives – like Queen Anne's Lace – carrots need to be isolated to keep them from crossing. Isolation is accomplished either with distance (0.5–3 km from relatives, depending on physical barriers) or by using methods like cages or bagging. Familiar carrots grow into large plants, creating hundreds of little white flowers atop branching umbels. I think they are quite pretty. Wasps seem to like them too, which is good since carrots are primarily insect-pollinated.

When the plants are dying back and maybe two thirds have seeds, you cut the stems, put the seeding umbels upside down in paper bags, and wait for them to totally dry out. Then you shake them, inside the bag, so the seeds fall out, and clean the seeds – maybe using a sieve, carefully using a blowdryer or fans to blow away the lighter chaff and then screening the seeds for size. For larger amounts tops might dry on tarps and cleaning be done with a system of buckets, fans/air compressor and screens. It depends; each saver has nuanced techniques.

When I ask savers which seeds they recommend to someone thinking about beginning to save seed, as you might imagine, they never suggest beginning with carrots – or any biennial for that matter. Tomatoes are popular despite their more involved 'wet method' (described in Chapter 2), but beans and peas are the most common suggestions. Beans and peas make saving easy – they are self-pollinating annuals and have obvious seed pods with big seeds (which are also food). Most consistently though, advice for beginners takes a generalised form (with some variation) of: start with something you love; watch and tend it carefully; have fun; store in dry, cool, dark places; be sure to label; and, do not plant all your seed – just in case. With these directives, savers imply the sensitivities and attachments that form through experience with seeds, hinting at processes of learning and training. These processes are made all the more clear when savers simplify their instructions further, paring them down to something like: be passionate, expect failure and 'you'll figure it out'.

Sensory Engagements

Seed savers describe sensuous engagements with the seeds themselves, as well as the various forms seeds take throughout their lives. Among the many shared stories of such experiences with seeds are evocative tales of:

cutting fingers on spiny spinach seed cases
jumping in surprise at the explosive pop of a wisteria seed pod
slightly spicy tastes of radish seed heads
gazing at the many colours and patterns of beans and corns
earthy smells, wrinkling skins and questing eyes of seed potatoes
watching pea tendrils reach for supports, and later enjoying peas cosied up in pods
discovering crinkly beet and insect-like calendula seeds
catching up with tiny, rolling brassica seeds
smelling fermenting tomato pulp
feeling bumpy, ordered rows of sunflower seed heads
contemplating alliums' architectural seed heads and curling stalks
fresh, sharp tastes of garlic, coriander and ginger
children playing with rattling poppy seed heads

Fig. 6.2   Calendula flower and seeds

Sensory engagements of saving are not only part of the practice, but sought out, providing one reason to continue saving seeds. In response to a question about how she decides which seeds to save, Hannah replies:

> It goes like this, and it has been like this since the ancestors discovered the art of saving seed: if you really like the taste of something, save the seed and plant it next spring and enjoy it again next summer. This also counts for plants that you find especially invigorating, like, say, stinging nettle.

Why save seed? In this case, to repeat a particular sensuous experience: a taste, smell, sight, sound or feeling. For Hannah, a good taste or a sense of energy inspires her to learn how to save those seeds but other sensations inspire as well. Paul, for instance, tells of *nicotiana*, one of his 'all-time favourite plants':

> we have an old variety, hardy, but a bit gangly; we have grown it so long I have lost the name of the variety, but I think it might have come from Dominion Seed in about 1970; I doubt it is in anyway unique. The flowers are varied, from white through to magenta, and it has a heavenly scent in the evening. At dusk as you approach the house and garden the most noticeable thing is the sweet perfume of *nicotiana*, which is bred out of the stocky modern plants. As a seed it is very easy to save, just shake an older plant into a paper bag. I save from several plants to make sure I get all the colour variations. In mild winters a few plants survive and flower next year. Volunteers frequently appear.

The smells and colours of Paul's *nicotiana* entice him to save their seed, despite his 'doubt that it is in anyway unique.' The sensations that come with the flowering plant might, in a mild year, go on without Paul's saving but he ensures their presence. His *nicotiana* saving is 'very easy', but careful.

One of the ways in which savers share their experiences, as Paul's story of *nicotiana* suggests, is by telling about their favourite seeds. I started asking people about their favourites after many casual conversations in which savers spontaneously said things like 'oh, this is one of my favourites', 'you'll like this, it's a favourite of' this or that person, or 'this is the best' something. I am glad I started asking explicitly. The stories and senses of saving offered in response connect me back to these savers, their gardens and fields, their loves and lived experience, and to the seeds – some of which became part of my own collection. Here are two savers' accounts of favourite tomatoes:

> Well, you've seen my garden. I have a lot of favourite seeds. I'd have to say that my two favourites are two local Mennonite heirlooms. An orange tomato, which is a Mennonite variety that has been grown in the Waterloo area for at least 90 years, close 100 years, and we know that it was brought here by Pennsylvania German

settlers before 1910. It may date back further in Pennsylvania, we don't know. But it's a great tomato. It's really thick flesh, not very juicy, but it makes a great tomato sandwich because it is very solid. And not very many seeds. We had pizza for dinner tonight made out of this orange tomato sauce. It looks a little weird, but there's very few seeds so it makes a great sauce. You just cook the tomatoes down, and you don't have to strain it. There aren't many seeds to speak of. Which is the down side of saving seeds from them, because you have to process more of them to get the seeds. But I'm not growing them in any quantity, so that doesn't matter. I just need a few for replanting and for sharing. (Denis)

I don't have a favourite – I love them all! I suppose my greatest favourites are those that have wonderful, rich stories behind them – I love the stories! One example would be 'Bull Heart', a large, heart-shaped tomato that has a meaty texture, a wonderful taste, and is a prolific producer of large fruit. I came by the seed from a regular customer at one of my seed shows. He came to me, and said he was getting old and he'd like to give me some seeds of a tomato that he had been saving for years. He had received them from an elderly Czech woman when he lived in northern Ontario decades ago. He told me that he was now getting old, and didn't garden so much, but he wanted to ensure that this tomato lived on. How could I turn that down? I planted the seeds, harvested the tomatoes and then planted those seeds the next year. I now offer the seeds in my catalogue, and it's one of my favourite sandwich tomatoes! (Jane)

Both Denis and Jane note that they do not have only one favourite. In truth, I don't think any saver does. I often got lists and multiple stories of favourites, each with a different set of reasons attached. In seed saving favourite is plural.

Seed materialities and saver senses feature in these favourite tomato stories as part of eating and saving experiences. Both Denis and Jane articulate strong appreciation of the taste of their tomatoes. It is clearly one of the main reasons to keep saving these seeds, particularly when other tomato seeds, plants and even tomatoes are relatively easily purchased. Senses – especially taste – frequently provide a reason for saving. Denis notes the 'really thick flesh, not very juicy' tomato with 'not very many seeds' made for great sandwiches as well as challenging practices of saving and sharing. Jane focuses more on the story of her favourite, but she is careful to describe the tomato itself as well: 'large, heart-shaped' with 'a meaty texture, a wonderful taste, and a prolific producer.' As this suggests, Denis and Jane value the sensations and seed-plant materialities of their favourite seeds. Considering saving particular tomatoes with Denis and Jane helps convey some of the sense of saving for these savers in which stories, materialities, ideas, embodiments and valuations combine.

With these tales of favourite tomatoes come some of the histories of the seeds themselves – one more personal, the other more historical, but both deeply connected with the seeds and why they are saved year after year. There is a combination of love of the story itself as well as respect for the seeds and savers of the past. Irene explains this deference further, stating: '[t]here's that sense of connection to everybody who has come first, who's also been part of the seed lineage. And I really love that about seed saving.' A connection to the past through seeds is part of why Irene loves saving, but this comes with responsibilities. She continues, 'I think that's where that sense of obligation and responsibility in me comes in. 'Cause I feel so beholden in some ways to take care of the heritage that's been passed on through all these other different savers and their customs.' For Denis, Jane and Irene, the seeds and their stories come with senses of connection and fascination as well as responsibility. These stories of saving favourites are stories of attached savers and seeds.

Living with Seeds

Seed saving is partly about growing plants, but it requires a shift in thinking. As indicated earlier in contemplating carrots, plants that have the right size, maturity, tastes, and so on, do not go to market or get eaten; instead, they keep growing and dying as they become seed. Inhabiting seed saving worlds, or enacting seed saving, involves living with seeds (and manifold others) as they live their own lives. Attuning to seeds and seed saving is, then, partially about recognition of the rhythms and cycles of seeds.

Savers highlight entanglement with seed cycles and worlds as key to experiential learning, and the pleasures of seed temporalities is one of many reasons why people enjoy saving. Comments like 'seed saving puts me more in touch with nature and the cycle of life' (Jane) and 'I wanted to be able to make a full circle with seeds' (Dara) pepper the interview and survey results. Anna explains:

> It feels like finally as a gardener, the experience is whole. That is, this year's garden has a continuity to the next year's – it seeds the future generations of plants. It seems to me that our culture is quite death phobic, and we tend to cut down plants as soon as they start to wilt or go to seed. Now, I feel I have learned to honour this cycle more and am often fascinated by how different plants change as they go to seed.

Saving seed requires connection with life cycles that include not only vigour but death, and which beget a renewed life through the seeds. For Anna, saving challenges gardening and wider cultural norms that hide or avoid death, and also offers opportunity for learning.

Participating in the entire lifecycle of seeds allows deeper understanding of seeds and makes the experience more meaningful. As Anna indicates above, the significance of seed saving, for some, exceeds experiences in the garden and the seeds themselves. Hannah concurs: 'I think that being attached to real cycles makes me think in terms of how things occur naturally and that to think in these terms makes life more valuable.' For both Hannah and Anna, learning with seeds' cycles leads to revaluing life – and death – in broader ways. Further, participating and valuing seeds' cycles contributes to an awareness that seed savers (and other people and things) are part of continuing worlds. Hannah continues: 'Saving seeds integrates me into a cycle that goes on forever and ever, like tides and full moons and all that. It is reassuring in the winter to know that as long as there are seeds waiting to be planted, that Spring is surely on its way.' Hannah feels personally reassured in the sense that she is prepared for the future season with her saved seeds, but she also connects with 'a cycle that goes on forever and ever.' This second aspect suggests an understanding of ongoing cycles, such that her relations are with worlds already in progress, worlds which affect her and with which she reconnects through saving. For Anna and Hannah, saving shifts senses and worlds – these savers are affected and respond.

Living with seed rhythms brings challenges. The spatial and temporal challenges associated with seed saving are the most cited challenges in the survey, an importance also revealed in interviews. The timing of seed saving can be quite important, and for some inconvenient. Commitments to family and work, limited experience and the intensity of seed harvesting can make getting to seeds at the appropriate time difficult – particularly when weather conditions shift quickly or are unfavourable. Additionally, providing the growing and storage space required to accommodate seeds' cycles is not always easy. Sometimes living with seeds can be too cosy when seeds take over rooms as they dry, or too overwhelming when they require immediate attention.

What people, seeds and places can do, dynamic as this is, is constrained by (among other things) the forms taken. As examples: if a person's dexterity is limited then planting small seeds or sorting seeds from chaff proves difficult; if a place is small or the soil poor then some seeds will not thrive; if seeds are not properly cared for while growing then many will not be able to sustain themselves for the next growing season. Physical limitations, sometimes real and sometimes imagined for the future, are often sources of frustration and/ or sadness when what was once doable, or what is thought should be possible, becomes unthinkable. Feelings of loss, real and potential, of abilities and collections are particularly evident with elderly persons although these issues are raised across ages. Interestingly, physical constraints can be both demotivating and motivating. Melanie, for instance, began growing from seed when she discovered she was very ill and wanted to better control what went into her body. She now continues saving seed to share knowledge with her children and engage in practices that she believes are beneficial for the wider world.

Aesthetic values and supporting technopolitical networks that favour particular stages of plant lives – positive when green and flowering, but negative when browning – present a different kind of challenge (see Robbins 2007). Seed saving – occurring as it does in the browning stages – becomes 'a problem' in neighbours' judgements that savers should 'clean up their yard'. Gail explains differences in living with seeds by comparing her growing with growing in other gardens:

> If you grow plants to maturity, you will be surrounded by the cycle of life, birth, growth, decay and death. Plants in your garden will dry on the vine, looking very untidy, flopping over or shading on surrounding plants, spreading seedlings where you might not wish them to grow, etc. In short, when you work with nature, you must let the plants grow more the way they do, rather than in a more controlled fashion. In my viewpoint it reflects the vibrancy and diversity of nature. In other people's eyes, it may resemble the 'untamed jungle'.

Aesthetic appreciation here links practically with the ways in which seeds lead their lives and the times and spaces that attend these cycles. How seeds take up space, continue living when other plant parts are dying, disperse themselves, perhaps becoming weeds, are all part of how Gail differentiates and grasps seed saving. In her saving, Gail accepts less control, finding great value in 'the vibrancy and diversity of nature' and believes this is reflected in her respecting the full cycle of seed to seed. She not only learns to see seeds and their differences, but becomes a different kind of grower as she does so. It is this kind of transformation both Latour and Haraway suggest in their formulations of learning with others, and it is one Gail clearly values.

Appreciating the cycles of seeds may spur conflicts among neighbours, but it can also be creative when conversations evolve around these differences. For Melanie, who decided to save seed in her front yard, conflicts arose in which neighbours were critical of the lack of lawn and the unkempt look. However, over time and with much discussion, this turned into acceptance and appreciation of her approach. She now gets visitors walking by the garden and chatting, has continuing friendships with neighbours interested in saving seed, and recognition of her skills and passion has fostered a community garden of native plants in which she plays a fundamental role. Though this is not necessarily what will occur with conflicting relations, it is a possibility raised through front yard seed saving.

Feeling Attachments

People and their practices hold together through mutual – though differential – feelings and attachments, with important implications for participation (see Ellis 2011; Lorimer 2008). People practice seed saving, at least in part, because they enjoy it – because it brings them satisfaction, joy, challenge, wonder, and so forth. As Alice puts it: 'I get a real thrill out of seeing seeds I have collected begin to germinate. In fact, sometimes I get so excited my husband laughs at me – the miracle of life and all that good stuff.' Will explains his own sensuous seed relations as a clarification of his interests in saving, saying:

> You know, it's not just the political or practical part of seed-saving, it's the sensuous and visual and tactile experience of collecting them and storing them and simply touching and looking at them. Seeds are a miracle and to be aware of how they grow and turn into their mature form to me is such a gas!

Will's statement makes connections between sensation, experience, learning, fascination and joy as well as offering a reminder that seed saving is political, practical and sensuous. As Will indicates, the practical and the affective are not separated, but neither are they reducible to each other; rather, they flow into one another as experience with seeds fuels curiosity and wonder within seed savers.

Much of a seed saver's experience with seeds occurs in growing spaces as seeds take form as plants, and a sense of wonder and attributed beauty to particular plants in their processes of becoming seeds was common among seed savers. Phil conveys these feelings in relation to lettuce (see Figure 6.3) as we walk through his farm:

> This is a lettuce patch. But of course, it doesn't look like your ground hugging lettuce. This is what happens when lettuce goes to seed. And you can see how different all the varieties are. They're quite spectacular – they're gorgeous. For me, just from the past few years really getting into it, I have over 100 different kinds of lettuces now, and they're all so beautiful as lettuce. But then when they are going to seed they don't have, you know, it's a different quality. And they're different in a different way again. And you don't know which one's going to be more like a candelabra and which one's going to be like some sea urchin. The first ones are just starting to flower. Little yellow flowers and then they become like dandelion fluff and then you just gotta shake it all into a bag and you get all the seed.

For Phil, the diversity among lettuces and their forms as they become seed are 'spectacular' and 'gorgeous'. And, in part, this kind of wonder led Phil to expand his saving practices. Phil does not only point out beauty and differences, he moves easily from form to sense to practical instruction in his effort to situate me in his lettuce seed patch and his seed saving. It should be clear here that saving does not offer a detached gaze (de Certeau 1984; Haraway 1991), but instead a multi-sensorial engagement. In this way, it reflects an approach similar to that offered by Wylie (2002) in which seeing is embodied, emplaced and intertwined with the world being seen. Sensing through seed saving reflects this more integrated understanding of sight, and connects sensation with wonder and attachment.

Fig. 6.3   Lettuces going to seed

The enchantment of savers with seeds and saving also connects with experiences with other nonhumans. In response to the survey, one seed saver tells such a story:

> The first time I picked petunia seeds, it was a warm late summer evening. I was sitting on a stool and working away when I heard a whir of wings. I sat very still and in the half light of evening, the white-lined sphinx moths were feeding off of the flowers. This was the first time I had ever seen them and I was enchanted. I continued to collect and had the moths actually land on my hands while I was picking the seed heads. A truly magical connection to my garden that day.

Through the sensate practice of petunia seed collection, this saver felt 'a truly magical connection' with a late summer garden, a warm night, moths and lingering flowers. Seeds here become background, but without the petunia seeds to be saved this saver would not have been out in the garden and would have missed the experience. Adam expressed similar sentiments about seed saving serving multiple engagements while we wandered through his growing space – a rented section of a larger farm. His appreciation flows into awareness of insects' desires combined with their practical roles, connects to his own learning process, and returns to aesthetic appreciation:

Here's a great example of a beautiful crop that you would never know! Any guesses? This is radishes! Isn't it gorgeous? These are radishes. Totally beautiful eh? Most people wouldn't get to see that radish is a beautiful flower. Yeah, just a tiny, just the tiniest little tinges in there [of pink]. Here's where a lot of the bees are obviously, there's quite a lot of bees in here right now, they're preferring the radishes at the moment. And the moths, moths are great pollinators as well. There's flies that come in here. You quickly learn what are great pollinators. Lots of flies in here if you look around. Yeah, so beautiful, you know, a beautiful stand.

After looking around, on what was a warm late summer's day, he broke open an early seed pod from one of the waist-high plants we were standing among so I could smell and feel the pod's just forming seeds. Adam hints, with his quizzing of my own knowing, that part of the thrill of saving is the seeming rarity of this kind of experience for others. But he is not attempting to protect this experiential knowing; instead, Adam invites me to share his experience, enthusiasm and passion. Though the particulars varied quite extensively, each trip to a seed saver's garden or farm brought this kind of welcome and enthusiasm with it. Seed saving, for these savers and seeds, becomes sensate and passionate practice.

Some seed savers extend their joy and satisfaction to other matters of concern. One surveyed saver, for instance, moves from the wonder of seeds, to satisfaction in work, and then to possibilities of self-reliance:

I started saving seed at the age of 5 (I'm now 48) by collecting nasturtium pods in my pockets as I played in my mother's garden. I was always captivated by seed pods as somehow magical. When I understood some years later that those pods could then be used to grow new plants – wow – blew my mind. I've been a seed saver and grower for many, many years and find it to be the most satisfying work and work that will be ever more important in the years ahead. I know that I will be able to share my wealth of seed knowledge and from my collection with others as our communities need to become more resilient and self-reliant on locally grown food.

Taking another tack on how seed saving connects to wider worlds of ethico-political concern, Irene explains:

I have this sense that the spirituality of many cultures is very much about their day-to-day life and relationship to the earth. And that's certainly how I see my own spirituality. And so the things that we're doing, like growing our own food and saving our own seed and keeping diversity and seed inventories on the planet, you know these are the day to day tasks that we do but they are also the things that we pray about and wish for and ask our deities for and stuff like that. So, it's kind of a place where the practical and spiritual connect.

This spirituality is not one distanced from everyday living. Rather, the embodied practices of planting, tending, selecting, storing and replanting seeds becomes part of a spirituality lived every day, and, in turn, spirituality becomes part of the practices of seed saving.

It would be easy to fall into a romantic notion of seed saving and all that it, and those involved, can do; however, seed saving is a set of living processes and a purely romantic version would be a sanitised one. In their investigation of efforts to create 'living cities' Hinchliffe and Whatmore (2006: 133) highlight the possibilities that emerge from engagements with socionatures, but remind us that different co-productions are not necessarily unproblematic:

> Let us underline again that talk of push and attachment should not be read as a return to an unproblematic sense of present presences. Nor should it be read as a romanticism of local becomings. This is a productive relationship whereby presence and absence can only form part of the story. The feelings and passions that are co-produced in these and other activities are tolerant of be-comings and goings, of likely presences, and are experimental in that no-one is quite sure of what will come of the attachment.

Hinchliffe and Whatmore are speaking of urban greening experiments here, but their statement is appropriate praise and caution for seed saving as well. Seed savers consider their practices overall positive – they continue engaging because they get joy from it and they find it interesting and satisfying. But, as suggested already in relation to challenges of living with seeds, seed saving (and growing more generally) is not always easy or pleasant.

The responsibility that people feel toward seeds, places and previous generations of savers can be empowering, but they can also become overwhelming. How does one justify continuing with one type of seed and leaving another? How can savers ensure that seed lives and stories continue when they are no longer able to act as savers? This last question is particularly concerning for many older savers who want to pass along the seeds they have been saving, but do not know someone interested. Several seed savers receive gifts from older seed savers imploring them to continue saving particular seeds that have passed through generations of families. In one such story shared with me, a saver who had written a seed saving article in a local paper was sent some pea seeds from an elderly man from another region who could no longer continue saving seed. Soon after, the man died, but this woman continues to save seeds from his peas. She has also returned some of the seeds saved through the years to a local museum in the man's region where the seeds now grow along with notations on their paths to and within Canada.

If things do not come together in particular ways and seeds die, savers often feel guilty – whether they had control over the results or not. This is especially the case if the seeds that have come to them were gifts, retained in memorandum or rare. Rachel tells of one instance in which a close friend asked for her help in saving garlic while the friend was away for a few years. This garlic had been saved over many years, moving with the friend to and from various locations. Rachel felt pressure to accept the task and succeed with the garlic, though she had not saved garlic before. She thought losing the garlic would mean not only the end of the garlic's living and evolving story, but tension in a longstanding friendship. Everything came together well (if not as planned) and the garlic continues to grow, now in several different places, but this is not always the case.

Another saver told of potatoes grown by her family for generations in France. Her father brought some of these potatoes to Canada when he immigrated and continued growing them. In his final years, however, he became ill and moved from the small farm where she had grown up with the potatoes. In the process of the move and her father's subsequent death, the potatoes were lost. This saver told me that a few years after her father's death she found similar potatoes, but they were not 'the right ones'. They did not carry within them her and her family's living memories; the right ones were gone. Pressures of seed saving are experienced also with the removal of varieties from a seed company catalogue, drastic changes in weather, tendencies in breeding toward hybrids and biotechnologies, changing legislation regarding property rights, and more. Each saver and seed, and relation among them, come with challenges and feelings of responsibility – welcome or otherwise. Despite the disempowerment that removed options and pressures can engender, savers generally use their frustrations and worries to fuel and reinforce, rather than abandon, their attachments with seeds.

Learning with nonhumans is, as indicated by Ellis (2011: 784), 'to achieve a wondrous (enchanted) encounter and to identify and see accurately', but it is also more than this. The learning that seed savers do with seeds not only combines joyful engagement with training, but that training serves to open different possibilities while shutting down others. Seeds and their savers thrive only as well as they have learned to live together, and some hard realities are faced along the way. Cultivated seeds would likely die without savers' care, though some would not. Some seeds die because savers have not yet learned to attend them, and others thrive despite attentions. Many savers' food – at least partially – comes from these seeds that affect our tastes and nutrition. When a saver selects seeds, s/he not only carries on a lineage with desirable capacities but discards those that fall short. As we have seen, the complexities of learning seed saving not only alter the participants, but change their collective possibilities.

Acquiring Skill

> Actually, that's what I've been doing lately. I'm walking out around my garden a
> lot just with a whole different eye, looking at my plants going – ok, where is the
> seed here, what can I collect, how do I do it, at what time? You know, just kinda
> looking at each plant. And so I think that's part of it. It's almost tuning me into my
> plants more. ... It's just a different way that I'm examining more closely. Instead of
> just their growth, and their flowering time, how big they get and all that. It's where
> do you start from originally. (Beth)

At the time of this conversation Beth considered herself a novice seed saver, and
was explaining to me how seed saving was different to the gardening she had
been doing for many years. In doing this, she details how to save seeds and save
them well requires walking, watching and inquiring differently. Her seed saving
eye is different from her gardening eye, though they may complement each
other. The questions she poses suggest the learning that comes with experience.
On a later visit to her garden, Beth laughed after noticing her actions while we
walked and chatted, explaining that her hands seem to automatically weed, head
and prop up plants without her mindful effort. As a longtime gardener, she had
embodied knowledge of when and how to do these things. Seed saving, as a new
activity, required conscious engagement and questioning to develop knowledge.
Beth's experience illustrates how seed savers expand sensitivity and sensibility,
developing various skills.

The sense-abilities and immersions of saving seeds then, are part of learning.
Doing seed saving well – finding 'the zone' as Haraway (2008) calls it – requires
ongoing, attentive engagements between savers and seeds. It takes, as one
surveyed saver put it, 'practice, practice, practice'.

Limited skills and experiences constrain not only what is done, but what is
imagined. If a saver has not learned about biennials, for instance, they might
not understand why their carrots or onions have not produced seed or fail to
imagine that a second season's growing might produce seed. Through experience
therefore, practices and imaginings alter. Phil attests to this through his saving
story about dried beans, recounting:

> I realised the reason most people don't have a very easy time with beans, is they
> find them hard to digest. All the storegoers are all, 'aw, beans, ugh!' because they
> don't know what it's like to eat fresh [dried] beans. All the beans in the stores,
> pretty much, have been sitting there for a long time, many for 3, 4, 5 years, and
> they oxidise and they lose a quality. You can see that. If you grow your own beans
> and you look at them within a year or so, they have this luminosity. They're just
> glowing. But if you look at most store beans, they're dull. They've lost their energy.
> It's the same when you eat them. When you cook up your own homegrown beans,

they don't give you indigestion, because they haven't oxidised and deteriorated and they cook up much, much faster. So, that's a story of my beans that was a total revelation. Now, beans could be a way, way, way, way better food than we imagined.

Phil's dried beans interfere with assumptions about bean eating and indigestion, becoming more than just a protein ingredient (or gas inducer) bought in a grocery store. By growing and eating saved beans, Phil learns how they affect people's digestive processes and how they are affected by storage in contemporary corporate agrifood arrangements. This experiential knowledge informs his ideas about growing food, reaffirms a commitment to seed saving, and fosters a desire to grow more beans and share his experiences with others. Overall, a different – and in his mind better – living with beans is revealed as Phil develops his saving and cooking skills.

Eventually, skills that once had to be learned become so obvious as to seem not worthy of mention or notice. Here, the regular practice of seed saving, as Hitchings and Jones (2004) suggest in relation to other plant practices, becomes so embodied as to become unthinking. Over the course of my participation in various training sessions, it became clear that questions from novice seed savers remind those more experienced of the learning process, and the particularities of saving. On one workshop's field walk for example, we came across very large, over-ripe, yellow cucumbers. Just as we passed the cucumbers, Martin, the workshop leader and a lifetime seed saver, was asked why the cucumbers remained in the field when most had been cleared away. He replied, 'Oh. I guess that looks strange. You let them get that yellow before harvesting for seed, otherwise, the seed isn't as strong for the next year.' He then removed one of the cucumbers from the vine, encouraging us to feel how it differed from a ripe cucumber ready for eating. To my hands, this cucumber was softer and yielding but was not yet squishy; to Martin's glance this cucumber was almost – but not quite – ready for seed saving.

Recognition and skills of experienced practitioners may seem wondrous to those less experienced. Ellis (2011: 772) discusses naturalists' species identification practices in a similar way, writing of how 'a virtuoso naturalist is to be able to transcend disciplined attention to detail, and to see a species accurately in a moment of flash recognition that apprehends it as part of an ecology of relationships.' In this case, Martin's learning had become innate, so he was able to recognise a seed-ready-cucumber in a way that we workshop participants were not. Ellis's exploration of identification by naturalists might be translated for seed saving as a sense of what a plant needs, when its qualities exceed those typical of its variety, or when seeds are ready – integrating seed capacities, saver expectations and cultivations, and ecological processes.

In sharing his knowledge it became necessary for Martin to revisit learning that had become part of him as a seed saver, reminding him – and the rest

Fig. 6.4   Learning through practice in the garden

of us – that this knowledge was not always so obvious. It is not a magic or mysterious knowledge; rather, it is knowing acquired over time as savers pursue their interests. A virtuoso – naturalist, musician or seed saver – requires more than just aptitude; they must have passion to continue and excel and skill in execution. But this does not detract from the wonders of such activities, instead enchantment and joy combine with meticulous attention and persistent learning.

Experimenting Experiences

Experimentation and curiosity play vital roles in saving seeds. It is through the experience of 'trial and error' that savers learn when seeds look ready to be harvested, how soil feels when it is the right time to sow, or how lettuce tastes when it starts to go to seed. Though many ways of learning were cited and combined, learning through experimentation – by trying things out in the field – dominated survey and interview responses to questions about how people learn to save seeds.

While saving experiments are not always successful in the sense of producing expected or desired results, they always contribute to seed saving. Brent relays this idea, echoing many others, by sharing his early experiences of saving:

> I began gardening as a young boy, helping Mom and Dad plant and take care of the garden. My father encouraged me wherever my curiosity led me. They gave me a science kit once that had corn seeds in it. I planted them – perhaps the first seeds I planted on my own – and all summer took care of the two plants that grew. Raccoons got the corn before we did, which was disappointing, but I don't remember feeling as if the 'experiment' had failed. Indeed it hasn't because my interest in growing things continued and evolved.

In this story curiosity, experimentation and care seem to have resulted in failure (though perhaps not for the raccoons), but Brent feels otherwise. Here, success becomes defined differently – as the continuation of interest and learning. Dara put it this way instead: 'I always say you can't make mistakes in gardening. You know, if it doesn't work out it's not a mistake, it's just an experience.' There is recognition in the words of these savers that encounters with seeds, however skilled savers are, do not always go as intended or expected. Savers agree, then, that experiments come with risk, failing only when they do not lead to further trials and learning (Latour 2004a: 196). And yet this insight requires qualification.

Different seed savers are variously able and/or willing to take care in what, when, where and how seeds and knowledge are put at risk. Radical failure in seed saving means seed death, an event significant in and of itself and one that comes with consequences beyond that loss. Experienced seed savers, including those who grow seed as primary income as well some who do not, limit the riskiness of experimenting. Precautionary measures include doing things like not planting all of the seed of a variety or running field trials in addition to primary saving activities. Learning well therefore, involves attentive engagements in experiments that risk material, sensory and cognitive outcomes (see Stengers 1997), but experimenting must be supported with alternatives for seeds and savers if things go badly wrong.

It is important to value the skills that are built through experience over time through seed saving, but this is not to suggest that savers feel the practice should be left to 'experts'. Quite the contrary. Lena explains how assumptions of needed expertise are generally unfounded when it comes to seed saving, at least most of the time, when she says '[i]t's like people are a bit intimidated – or think it's something that you need more knowledge for. So, it's nice to explain that in most cases it's really quite easy to do.' Answering a question about what she would tell someone just beginning to save seeds, Pat urges and reassures, 'Just do it! Experiment, play, There's no right or wrong', and then cautions, 'but be sure they're dry, avoid packing away anything with some moisture in it that may become mouldy which is always a big disappointment.' Irene too is both encouraging and cautious as she explains what she might tell a first time seed saver:

> It only takes one really successful seed crop to hook you ... Part of me would say just try it and have fun and don't worry about the details. Yet, I might not say that without also saying that seeds are really important and people have worked incredibly hard and incredibly devotedly to keep seed lines pure, so just to have that responsibility to honour that heritage.

Irene attempts to balance her enthusiasm for seeds and saving as well as the ease and fun of doing it, with a responsibility to the seeds and to their shared histories with people. Finding ways of keeping continuity of seeds and tradition, while also moving forward with learning and experimenting is a challenge for each seed saver and to seed saving as a collective activity.

Along with savers' encouraging words come warnings: Lena clarifies that 'in most cases it's ... easy'; Pat cautions of the dangers of mould; and Irene reminds of the importance of seeds and tradition, insisting on savers' 'responsibility to honour that heritage.' Many other cautions – each a reflection of a seed saver's own experiences – are offered, often regarding isolation distances, labelling and storage. Recent increases in the numbers of small seed companies have also highlighted concerns about quality and quantity, resulting in multiple efforts to provide training. Experimentation, while often fun and not always difficult, involves careful methods and skilful practice to attain useful results – as seeds and/or learning. Despite the cautions and caveats though, savers resoundingly welcome and cheer on those new to the practice.

Tension exists, therefore, between seed saving as easy and fun versus doing it well requiring respectful skill. Haraway (2008) discusses this complexity by reflecting on training with her dog, Cayenne. She argues that both she and Cayenne have altered through the training – in positive and negative ways. On the whole, she argues, training serves the interests of trainer and trainee in its complex combinations: 'training requires calculation, method, discipline, science, but training is for opening up what is not known to be possible, but might be, for all the interacting partners' (Haraway 2008: 223). In this view, the tension between fun and skilled practices is not necessarily a contradiction. As Haraway and the savers quoted in this section suggest, this tension can be a productive – bringing inspiration together with training, emboldening novices while encouraging skilful practice, remaining cognisant of the possibilities that come with learning and training.

Learning Collectives and Collaboration

In my retelling of marigold saving at the opening of the chapter I received instruction from a more competent seed saver, and then worked out my own methods in negotiation with my body, marigolds and an inclined growing space.

In this everyday practicing of learning to save marigold seeds, I was participating in the continuing processes of seeds, weathers, lands, and the saver I was helping. Lave and Wenger (1991) theorise learning as a situated, social activity in which people participate in communities of practitioners, forming relations with other learners – some more experienced, some less so. Instead of considering learning as an isolated activity in which a person acquires knowledge, learning becomes a social activity in which people, practices, and worlds are co-constituted. Further, in later work Lave (2009: 201) argues that 'participation in everyday life may be thought of as a process of changing understanding in practice, that is, as learning.' This reconceptualisation sits easily with learning to save seed as growers rely on other growers and various resources to inform their repeated activities, which are situated with seeds and spaces of growing and storage – just as in my first experience with marigold seed saving.

Lave and Wenger maintain a division among things and people in which people are the agents, which reflects a commonly articulated view of agency but one which is increasingly disputed and which I challenge throughout this book. I'm interested, instead, in learning as a collaborative coming together of diverse texts, devices, creatures, skills, environments and people – each of which make differences to the saving practices. Learning is an ongoing collective enterprise.

Of the multitude of ways that this thought might be explored, I am going to focus briefly on two sets of relations through which more-than-human collective learning occurs in seed saving. First, retaining a focus on human relations, I consider how other savers become part of seed saving activities. Second, pushing toward considering nonhuman agency, I look at how some things – most especially texts – become part of seed saving's learning collectives.

Learning with Other Savers

Conversations over the fence, at the kitchen table, at seed fairs, in the garden or on a field day make important contributions to savers' learning. This sharing of experiences and experiments also sometimes comes with exchanges of seeds, further influencing future saving. While some of the savers I spoke with had degrees in genetics or biology, all mentioned learning from other savers' experiences and in less structured environments. It is common to hear savers say things like 'that's the way my mom did it', 'my neighbour said', 'I went to [an event] and we talked about how to do it.' In fact, it is difficult to find a seed saver who has not learned through and with other people, whether these people are family, friends, workmates, neighbours, co-participants in workshops or authors. Over 77 per cent of the savers participating in the survey indicated that they learn through informal conversations with other savers, and when you add this to those who included workshops (42 per cent) and farm/garden visits

(41 per cent) the role of face-to-face communication and experiential learning is revealed as indispensable.

The two stories below come from survey respondents, each inspired by collaboration with other savers but as part of different kinds of relation – one as a worker on a farm, and the other as a facilitator of a local seed saving group.

> I started saving seed because I was doing a farm internship and they were growing the 'striped cavern' tomato. It was so beautiful that I asked them to show me how to save the seed so that I could grow it back home. I saved those and 10 other varieties. When I returned home, the following spring, I planted the seeds, not really sure what would happen. When they sprouted, I was admittedly somewhat amazed, even more so when they produced striped caverns. I was hooked. Eight years later I have my own tomato seed business and over 70 varieties of tomatoes!

> After realising that there was a need for education among the group of people who were trying to start up a seed sanctuary, I invited those who were interested to join me in a seed saving study group with me as facilitator. I do the preparatory studying and present the information and we all discuss, ask questions and share whatever other information we have about the particular plant family we are studying each session. There was a good response and we are now an engaged and dedicated group; our goals are to achieve seed saving mastery for edible crops and to freely share the knowledge we acquire.

In both of these stories, the savers speak of learning with other savers, relying at least partially on others' ideas and experiences to gain new ideas. Learning is put to use in these savers' own saving practices and tested through experience. Each employs the combined learning as they explore seed saving possibilities – one building a business inspired by a tomato seed variety and those people she had worked with, and the other sharing collective goals of learning to share knowledge and seeds through a local seed project.

Recalling the passion and knowledge of family members or friends who saved seed is common among savers. Reflecting on the path taken to seed saving, one seed saver surveyed tells a familiar, though singular, story:

> I grew up close to my grandmother who had a large vegetable garden. I never saw a commercial seed package at her house, she had a box of seeds she saved every year (all well labeled and with hand drawn pictures on the envelopes), grew all her own transplants, and come planting time always exchanged seeds and plants with all the neighbours who grew a garden. Thus, when I started gardening, saving my own seed where practical was a matter of course for me and I think this kind of saving and sharing of seed should be encouraged.

This saver pursued seed saving as 'a matter of course' based on childhood experiences with her grandmother's saving. Many savers continue to save particular seed varieties tying seeds with important people in their lives or stories they find intriguing. Continuity also takes form as ongoing practices, as illustrated here. Remembrances can be an important part of seed saving, and in this case, as with many others, 'as things are done, other events are remembered and re-placed into the present' (Crouch 2003b: 1955).

Sharing seeds and related knowledge is an important part of how seed saving gets done and how it continues. As Carolyn explains: 'Meeting with other like-minded individuals helps with motivation and education, as well providing opportunities for sharing seed and ideas. It helps to create a framework within which the seed saving and sharing process can take place.' Sarah, a small seed company co-owner and seed saver, relays her own experience this way:

> Well, I think we have a very strong community of organic farmers. Many of us do save seeds and many of us, we end up sharing seeds all the time. There's so much information about how to save seed. We talk to each other about how to save, and encourage each other. I think saving seed is harder than growing vegetables because you grow so many different seeds, it's like a promise and you don't really know what it is. There's a real practical knowledge and so much to learn, you can't learn it all by yourself. You really need other people to learn with you.

By sharing their experiences and their seeds, Sarah suggests that she and these growers are building stronger seed networks and communities. She differentiates between those who save and grow seed from those who do not – 'seed saving is harder than growing vegetables' – indicating that different knowledges and skills are necessary. Further, the complexity of learning seed saving drives her efforts to participate in a community of practice. In the views of these savers, connecting through sharing seeds and knowledges serves as a means to contribute to communities – their own and others. It is also a way of finding support for and ways of improving practices. Learning to save seed enrolls more than just one saver.

Learning with Other Things

In learning to save seed, the reading of texts combines with people's telling and listening to stories and in-field practices. Seed savers depend on various texts – books, blogs, websites, magazines and their own logs – to aid their saving. Internet sources are increasingly important, with blogs and websites being cited by 63 per cent of survey respondents as useful in their learning. But books still prevail (73 per cent of survey respondents regularly use them in their saving).

Practiced knowledge does not travel particularly well, and embodied skills in addition to responsive collaborating objects are just as (in some cases more) relevant as representative texts (see Carolan 2011: 142; de Laet and Mol 2000). This is an insight that seed savers know well, but this does not belittle the importance of texts. Rather than relying only on written texts, and the many contradictions and omissions that come with them, ideas are gleaned and then tested by seeds and savers in the field. Paul explains his learning process this way:

> Basically, I have seen others do it many times (and in different ways) and without any formal lessons, evolved my own methods. The books that I own or have read offer advice on how and when to save seeds for various species, whether they need to be scarified or stratified, how long germination might take, whether light is needed, what temperature or moisture level. However I am quite willing to experiment, just to meet the challenge of getting seeds to germinate.

This statement suggests that Paul, in ways similar to other savers, acts as part of a collection of humans and nonhumans in his learning process, taking texts' and people's advice and exploring these through his own experiments with seeds.

Texts participate in their own ways in seed saving. With regard to the role of texts, Hinchliffe et al. (2005: 648) observe that their use of field guides in a water vole field study added to understandings by serving as 'sensitising devices, diagrams that made water voles more rather than less real for those who started to use them.' In this view, texts are active participants in learning processes, making water voles more present for those studying them. In learning to save seed, texts serve not as knowledge to be acquired, but as sensitising artifacts that expand seed saving options and act in collaboration with learning, quite literally, in the field. Denis explains one of his early seed saving lessons:

> I saved a whole bunch of tomato seeds, and then I planted them the next year, and they were a whole mixture of different things. So I realised that I had done it wrong and I tried it again. I said, 'ok, when the book says that they'll cross when they're 5 feet apart, I guess that means they cross when they're 5 feet apart. Look at that!' I got some crazy mixture of varieties in that row, which was kind of interesting. I learned by doing.

Denis, through reading about tomato seeds and crosspollination, becomes more able to understand how those tomato seeds came to be so mixed. As he says, he 'learned by doing' but also by adding to his practices through reading, and rereading. Further, by sharing his story with me, Denis includes himself in my – and now your – collective learning about seed saving.

In many ways, seed saving involves adapting and using what is at hand. Seed savers frequently reuse various technologies meant for other purposes: cake

pans and window screens become stackable drying surfaces; kitchen sieves, air
compressors, standing fans and food processors are seed cleaners; cotton/linen
pillow cases hanging from rafters serve as animal/pest resistance mechanisms.
Technologies doing other work also get enrolled – freezers stratify and fridges
store, their waste heat sometimes helping other seeds germinate. Savers also
invest in new items such as germination mats, grow lights, row covers and
greenhouses, or rebuild things such as hand-cranked seed cleaners and threshing
boxes. In various ways people, seeds and artefacts come together in practice, as
one surveyed saver illustrates:

> I learned a great technique of seed saving from [an experienced saver]... where you
> first let your healthiest plant of a particular variety go to seed, cut it, then shake
> the plant over a screen with a sheet underneath to collect the seeds, then finally
> put the seeds into a stainless steel large bowl and gently use a blowdryer to blow
> out any remaining debris in a circular motion as to not blow out the seeds.

This explanation of a particular technique reveals a collective of participants –
savers, seeds, plant remains, along with blow dryer, bowl, screen and cutters –
working in collaborative ways – gently circulating air, shaking, sifting and
collecting – to achieve saved seed.

All seed savers engage with technologies in their saving but the numbers,
purposes, kinds and valuations differ for each saver, seed and growing space.
Plastic, for instance, is a controversial in whatever form it takes. In storing
seed collections (Figure 6.5), some savers use old film canisters, margarine or
peanut butter tubs, or 'Tupperware' containers. However, others feel that these
containers are not ideal since plastics may release chemicals that negatively affect
the seeds' present and/or future being. These savers choose, instead, to use glass,
paper and cloth to store seed. Savers may also be unaffected by such debates and
employ plastic and/or non-plastic containers without concern.

Fig. 6.5   Stored collections of seed savers

All of the technologies mentioned here – among many others – reshape saving possibilities by expanding, though also constraining, the ways in which people and seeds relate. As I suggest in Chapter 2 using the example of tomato seed saving, artefacts do important work in practice. So the observation that texts serve to sensitise in ways that may expand possibilities can be extended to many objects that feature in saving practices; further, these objects 'have a certain effectivity of their own' and alter the practices that they join (Bennett 2010: xvi). Some things are more fundamental to saving than others – while a saver could do without plastic containers, they require containers of some kind – and, in this case, all present themselves as part of this practice in the service of seeds being saved. The saver quoted earlier in this subsection might not have looked at her hairdryer, for instance, as a tool of seed saving until learning of another saver's technique – but it now shows up differently, and this may prove useful the next time lettuce seed needs cleaning.

I have learned saving, like many others savers, through my own trial and error experimentation; reading various books, pamphlets, websites and blogs; speaking with many other seed savers about what they do, what goes wrong, what they plan on doing next; and attending workshops, field days and working with other savers in their own spaces. Each of the people and things that I learn with appear with me as I figure things out with seeds – saving marigold seed in my own garden years later, my first experience, recalled at the chapter's opening, still evokes and informs. Those I have learned despite of also make their influence known – a poorly researched book, a slow internet connection, a misunderstood conversation, a hybrid seed. There are also gaps, or silences – trials that did not work out or were not attempted, things forgotten or not thought of. I try to pay close attention to the plants and seeds, but I am still pretty inexperienced. I get it not quite right much of the time, but I am learning.

Summing Up

Saving seed is different from gardening, and it offers different opportunities for learning and relating. As people save seed they inhabit a world with seeds, learning to understand and share in the histories and capacities of particular seeds. Living and working with seeds connects savers, if only partially, with the seeds, as well as wider ecological cycles and landscapes. The knowledge-in-practice of seed savers is also multi-sensorial and affective, informed by experience and prior knowledge, requiring and building broader understandings of surroundings. Seed saving is experiential, and improved through skill, but surprise and unexpected outcomes are ever-present. Paying attention to seeds and their living both rewards and challenges savers. Learning seed saving is a collective endeavour involving seed, people, spaces, times and other involvements.

Learning or training opens up possibilities for seeds and savers, and their practices together, not only in the sense that saving may become more embodied, deeply known or wonder-filled but also in that it may propel, or be initiated by, ethico-political concern. There are hints in this chapter about whether and how seed saving practices become ethico-political engagements in the stories of saving as collective learning, as revaluation of life and death, as care-full attending. In the next chapter seed saving as collaborative ethico-political praxis is further investigated.

Chapter 7
Reconstituting Worlds Together

> If the world exists for us as 'nature', this designates a kind of relationship, an achievement among many actors, not all of them human, not all of them organic, not all of them technological. ... The commonplace nature I seek, a public culture, has many houses with many inhabitants which-who can refigure the earth.
>
> Donna Haraway, *The Promises of Monsters*

Cultivated seeds are living beings with their own lives, but they are also artefacts of human processes. Humans too live beyond, but also with the cultivated seeds upon which they rely. Here limited human control, or what Haraway (2008) calls degrees of freedom, is undeniable. By admitting our mutual reliance, and the heritage embodied in this reliance, the relations between seeds and people gain a significance that is not always recognised. Haraway, quoted above, draws our attention to this idea that multiple actors – some human and some nonhuman – reconstitute worlds together. However willing social sciences have been to integrate this kind of approach to human–nonhuman relations, it can still be controversial when it comes to some nonhumans – including seeds.

Saving seed is always ethico-political practice in the sense that seed and savers, in their intimate relations, make differences for each other and related others, but saving is also more than this. Acting as part of larger socionatural assemblies, saving intervenes. Seed saving presents possibilities for worldly engagement, but it is only through taking up these opportunities that worlds and participants can be changed. And it is only in the particulars that we discover whether and how relations become meaningful.

In this chapter I consider seed saving as ethico-political. In particular, two ways in which savers value their practices as contributing to better collective futures are examined: food provisioning and conserving biodiversity. In addition, this chapter explores the relative agencies of savers and seeds – their degrees of freedom in seed saving – in practices of selecting possibilities, adapting capacities, demanding accommodations and presenting opportunities. Taken together, these two sections illustrate some of the ways in which seed saving reconstitutes meaningful practical engagements. But first, a more conceptual discussion about active participation and agency, about ethical engagements and valuations is necessary. In this first section, I reflect on two approaches that offer insight into how savers and seeds act together – first, a version of stewardship and second, a feminist ethics of companion species. I argue that seed saving may be partially understood through stewardship valuations, but that more suggestive are approaches decentring humans and taking nonhumans seriously as agents with their own lives. The worldly engagements of seed saving are, therefore, collaborations – not always easy, equal or enchanting, but collaborative engagements none-the-less.

Acting In and With Natures

Saving seeds affords possibilities, but it is only through taking action together that seeds and savers may improve their worlds. People not only come to know the world through enacting it with others, but ethics is based in everyday experiences rather than abstract notions of right and wrong (Barad 2007; Varela 1999). We are all implicated – savers, seeds, researchers, eaters, pollinators and others besides. Through embodying, embedding, skillful everyday practices (see Chapter 6), people and seeds may recreate meaningful relations. The meaningfulness, or valuation, of these experiences is key for people to continue taking up 'challenging, skilful, sometime tedious activities required to keep something of value alive' (Higgs 2003: 244). As already indicated, it is not enough to be open to possibilities, though this is necessary; we must do something, on a regular basis, about those possibilities if we hope to sustain them.

The question of how to act within the world while respecting seeds and other nonhumans is a question that savers face on a regular basis. Should I water or not? Should I fence the space to keep out the deer, maybe they could be distracted with offerings elsewhere, or should their munching just be accepted? Should I plant this perennial here or there, in the company of this or that plant? Are these seeds ready for harvest or should I wait a little longer; if I wait will the weather hold? Are the seeds dry enough, or will they sprout, or mould?

Compounding concerns about taking appropriate action is anxiety about taking action at all. This worry reflects established valuations of natures as well as a more general sense of possibly destructive actions of humans within the world. For some savers, there is disjuncture between their practices and much naturalist and conservationist reflection, particularly the 'hands-off' approach; many agree that wilderness deserves conserving and that a weed is simply 'a plant whose virtues have not yet been discovered' (Emerson 1878), but this does little to address practical questions. In what ways is the plant, and its seed, out-of-place, and what should, if anything, be done about it?

Stewarding Resources

The question of how to act in natures is directly addressed in ethics scholarship regarding stewardship. In general these days, stewardship refers to responsibly managing natural resources – careful planning, efficient use and waste reduction are all part of the approach. As an advocate of agrarianism – one part of which is stewardship – as a way of living that opposes and challenges industrial and corporatising agrifood, Berry (2002: 39) writes:

> [t]here is another way to live and think: it's called agrarianism. It is not so much a philosophy as a practice, an attitude, a loyalty, and a passion – all based in a close connection with the land. It results in a sound local economy in which producers and consumers are neighbors and in which nature herself becomes the standard for work and production.

This conception contains some problems, not least its romanticism and insinuation that nature is a unitary 'her'. However, there is ethical consideration of nonhumans – lands, waters and animals – such that nature provides foundation for human activity. Further, agrarianism disrupts ideas that ethical treatment of nature requires 'leaving it alone' or that one should feel only aesthetic appreciation of nature.

These arguments rest squarely on the idea that human relations with natures, even and perhaps especially, productive ones, need to be recognised as valuable and ethical. Berry writes of agrarianism that it 'can never become abstract because it has to be practiced to exist' and that '[i]n order to be good you have to know how – and this knowing is vast, complex, humble and humbling; it is of the mind and of the hands, of neither alone' (2002: 43). Acting in nature, then, can be ethical practice. Further, it is through acting in nature that ethics takes shape – knowing how to be good 'is of the mind and of the hands.' This kind of practiced ethics in nature sits easily with savers who see their practices as meaningful engagements that yield food, conserve biodiversity, require skill and build local resilience.

Yet stewardship – some versions more than others – leaves in place both instrumentalism and human exceptionalism. In this view, a good grower (or consumer) stewards resources, otherwise their and/or other human futures are jeopardised. In this sense, stewardship takes an ethical position of 'enlightened self-interest' in which natures are valued, but only for their uses (past, present or future) (Thompson 1994: esp. ch.4). Here it is not only that human–nonhuman relations may be productive, but it is only in their productivity that they become valued. In addition, stewardship maintains an underlying assumption that humans control outcomes – that people can, in fact, adequately steward resources. There is, in this belief, a continuing sense of human mastery of nature – that nonhumans are both knowable and manageable (Palmer 2006). Natures are valued in some ways through stewardship that might check unsustainable activities, but humans remain separate from nature and the only agentic subjects. Light (2007: 9) suggests, in his own construction of stewardship as participatory ecological restoration, that this is not necessarily problematic; one need not 'see nature in and of itself as some kind of agent in and of itself' in order to 'care about the land around us.' To oversimplify, for stewards the point is to manage natural resources well.

This chapter will show that savers articulate stewardship values as part of their reasoning for participating in seed saving – combining instrumental valuations of seeds and natures with careful practice – but many savers do not stop there. Seeds do more than just constrain human actions or serve human purposes. Further, the labours and skills of saving are not solo endeavours. Necessary or otherwise, saving as ethico-political practice pushes beyond stewardship.

Jostling Stewardship Sideways

Plumwood, an eco-feminist philosopher, challenges dualistic and hierarchical approaches to natures. Plumwood's thinking is famously influenced by her own experiences – particularly her survival of a crocodile attack while canoeing in 1985. In retelling the incident she relays her shock at becoming food and her later insight that '[w]e are edible, but we are also much more than edible' and that all creatures – human and non – 'can make the same claim' (2002a: np). I am interested here in two aspects of Plumwood's insight: the decentring of humans; and the acceptance of instrumental use as part of ethical relations among organisms. Conflicts, competitions and deaths are part of how we live. For Plumwood, her crocodile-initiated near-death experience demonstrates this point, and rather than shutting out natures or demanding their mastery, the entanglement drives her to insist upon agency of natures.

Plumwood argues that human-centredness 'which includes the hyperseparation of humans as a species and the reduction of non-humans to their usefulness to humans,' serves neither humans nor nonhumans well (2009: 117). Through human-centredness not only do we (humans) lose a sense of the dynamism and liveliness of nature, but we also lose a sense of our own inter-relations and dependencies. As such, Plumwood argues, we threaten both natures and ourselves. Instead, she urges undertaking a dual project of '(re)situating humans in ecological terms and non-humans in ethical terms' (2002b: 8–9). In this struggle to rethink our world and our place within it we must 'work out new ways to live with the earth, to rework ourselves ... [to] go onwards in a different mode of humanity' (2007: 1) and become *'open to experiences of nature as powerful, agentic, and creative'* (2009: 126, emphasis in original). Humans do make differences, however, our knowing and influence are partial; in recognising this the hope is that we take our responsibilities all the more seriously (see also Clark 2011). For Plumwood, decentring humans is not an abstract or unnecessary project; it offers opportunities to rethink and reconstitute worlds in better ways. Plumwood's work is a call for action.

Since Plumwood's early work a diverse social studies literature endorsing non-dualistic and dynamic accounts of human-nonhuman relations has developed. Haraway's (2003, 2008) figuration of companion species notably inspires and provokes rethinking human-nonhuman ethico-political relations –

including, here, seed saving. In this recent work Haraway shares her own intimate encounters, particularly those with her dog Cayenne, to illustrate the opportunities and complexities of enacting response-able human-nonhuman relations. In this work Haraway retains organisms as her unit of analysis such that organisms retain some individuality but are not unitary, settled, isolated or impermeable. Haraway's companion species, from dogs to intestinal flora, are 'about the inescapable, contradictory story of relationships – co-constitutive relationships in which none of the partners pre-exist the relating, and the relating is never done once and for all' (2003: 12).

The point of studying such relations is not only to describe complex worlds of human-nonhuman relation, but to 'become worldly and respond' (Haraway 2008: 41). Learning to be affected (discussed in Chapter 6) is part of this, but we must engage further to take up these possibilities and ensure our worlds are made better by our doing so. Haraway explains the connections of learning and enacting better worlds through an example of chickens:

> A fundamental principle for me in animalhuman worlds is that we had better know more at the end of the day than we did in the morning if we consider ourselves to be serious about multispecies flourishing, or maybe just survival. It matters to know now that chickens recognize multiple social partners individually; it matters that they are subjects of their own lives in ways people used to reserve for themselves. ... it matters to show as fact – and to draw out the consequences in law, technology, and human labor training – that chickens killed in an argon gas mixture pass out without signs of pain for evolutionary reasons to do with the history of earth's atmosphere, but show multiple signs of acute visceral terror if gassed in a carbon dioxide system. Response to and respect for chickens requires at the very least knowing these facts effectively and making them unavoidable in daily life, both intimately and publicly. (Haraway in Potts & Haraway 2010: 326–7)

Haraway illustrates the importance of learning as caught up in techno-political processes of agrifood and further, that response-ability develops through efforts combining intimate and public engagements.

Haraway's ethico-politics, then, involves 'recognition that one cannot *know* the other or the self, but must ask in respect for all time who and what are emerging in relationship' (2003: 50). In other words, we must ask not only how we relate, but what our relations are doing to ourselves and others – how are realities reshaping through our practices together, and are we better off? Building on the previous chapter, this chapter and the next begin to consider how this kind of approach might connect with seed saving.

Seed saving, with Plumwood and Haraway, allows complicated, lively and partial accounts of practice and practitioners. Seeds and savers have long-standing and ongoing relations from which neither can be fully extricated; each needs and changes the other as they work together in seed saving. But seeds are not dogs, chickens or crocodiles. Seeds have their own capacities and tendencies, lives and histories, that condition and invite, are ordered and trained. Seeds[1] live and act, but perhaps not in forms as easily recognised by people. Moreover, seeds and their savers share degrees of freedom that shift with different phases and purposes of saving seed. Selection and growing, for instance, are differently uneven favouring interventions by savers and seeds respectively. Saving seed well means seeds and savers learn and adjust with each other, working to enact (hopefully) better shared worlds. Seed saving may therefore be a means for savers to produce good food or conserve biodiversity, for instance, but it is always more besides.

Meaningful Engagements

Seed savers begin and continue saving seeds for multiple and differing reasons. However, savers in the interviews and survey cite producing food and conserving diversity, along with building self-reliance as key purposes. Of course, emphasis varies by person and with how, where, when and with whom savers communicate their reasons. All interviewees except one expressed dedication to conserving cultivated biodiversity – genetic, species, ecosystemic – as one of several reasons that they save seeds, along with a belief that in general, growing and saving seed from heritage varieties is a productive way to contribute to conservation goals. All but one interviewee (not the same one) also included food production in their stories and reasoning. In the survey, food security and maintaining diversity were most commonly listed first.[2]

Though suggestive, these numbers do not really convey how seed saving is an enactment of (among other things) reproducing foods or reconstituting biodiversity; for that I rely on savers' accounts. In these accounts of meaningful worldly engagements, instrumental valuations manifest – after all, food sustains people and conserving diversity provides one means of securing future human food security. But even these practices become more than careful resource management when closely examined.

---

1   Here I need to clarify, as noted in Chapter 1, that seed saving involves not only 'seeds' as defined biologically (though these dominate), but a broader category of propagators including seeds, cuttings, clones, and so on.

2   Survey participants were given five blanks in which to list their responses to the question: 'For you, what are the most important reasons to save seed?' The top three of responses listed first were: food security; biodiversity conservation; and self-reliance. Overall – totalling responses from all five blanks offered – the top five reasons related to: self-reliance; food security; cost savings; biodiversity conservation; and local adaptation of seed.

Food Provisioning

Saving is not an isolated practice; rather it exists with other practices – including food provisioning. Seeds are not only the first link in agrifood production, they are often foods themselves. In both of these ways, they contribute to savers' efforts to provision themselves, their households, and their communities in ways that are healthy, safe, secure and tasty. Sandra explains her understanding of taste in relation to her saved seeds and improved (hybrid) varieties: '[t]he seed saving I have done showed me that my "old" stock is a bit better than the recent ones. Seems to have more punch, taste.' The tomato stories shared by Denis and Jane in the previous chapter similarly reflect on tastes of seeds that may not be easily accessible. Rachel furthers this argument by linking saving to healthy food, saying that she loves the taste of the tomatoes she grows each year, and that 'the quality of heritage foods is better.' Hannah pushes these connections further, explaining, 'I have tomatoes, and garlic, and beans, which I really like to eat and always can because I control keeping them available.' For these savers therefore, saving brings together complex food and seed worlds featuring stories, senses, materialities, valuations, histories and possibilities of what they understand as better worlds – better tastes, health, self-reliance, and more.

Fig. 7.1   **Peas for eating grown from saved seed**

Many savers combine food and seed concerns through their practices. Richard, for instance, cooperatively operates a small seed company that includes a community-shared agriculture (CSA) program and food provisioning for those on-farm. He argues that the 'best situation would be to grow ourselves a good part of our food. Food quality cannot equate uniformity. The uniformity of supermarket vegetables reflects the lack of genetic diversity of the varieties used. The more than 50 varieties of tomatoes we offer in our CSA baskets are NOT uniform.' While Richard speaks here more as a producer building diverse economies and food varieties, Lena explains her position more as a consumer who grows some of her own food as part of seed saving:

> I like the freshness of homegrown food and I like the idea of being able to grow food without pesticides. And that to me is very important. I've always felt healthier in the summer and fall months, when I'm eating everything out of the garden. I know it's better for me. So, that's one reason. And also it's just the idea of being able to grow your own food, not having to go and pay someone else for it. It's a self-sufficiency thing I guess. It gives me a lot of satisfaction.

Here Lena brings into connection eating food that is 'better for her' – healthy, homegrown, fresh, without pesticides – with a sense of satisfaction about increased self-reliance and avoidance of purchasing food. These statements might have come from any grower who includes food provisioning in their practices, and in some ways saving might be read as just part of this. However, savers suggest that saving itself brings agrifood possibilities.

One way in which saving provides opportunities is to open up personal food practices. Irene and Andrew each explain the possibilities raised by seed saving to consider food seeds and consume differently:

> Of course, in growing it, you end up also eating it, and coming to appreciate the plant as a food plant not just as a seed plant. But I would say that there are some foods that we grow, that I probably wouldn't have chosen to grow as food, I was interested in growing for seed. Particularly some of the beans. (Irene)

> I didn't place any value on using shell beans out of the garden until I started to save seeds and I started to be aware of the seed saving phase for dried beans. It kind of lead me to the realisation that I could grow shelling beans in the garden and it wouldn't be too much work in terms of collecting and saving the crop and from there to incorporate dried beans in the repertoire of things that we cooked. (Andrew)

Irene and Andrew each credit seed saving with exposing the possibilities of saving and eating beans. As it turns out, Irene appreciates but remains less interested in beans as food. Andrew, in contrast, seems excited by the change in his diet, cooking practices and tastes that come with saving beans.

Seed saving awareness also broadens agrifood understandings about labour and the agrifood order more generally. Carolyn explains that seed saving 'has also made me much more aware of the amount of labour and expertise that growing one's own food requires. I have always disliked wasting food, and now have a greater appreciation of the value of food.' By doing some of the work of food growing and seed saving herself, Carolyn has developed a respect for those who farm – their labour and expertise – and respects their (and her own) contributions by not wasting food. But Susan thinks more is necessary. She argues that it is 'imperative that we pay more attention to meeting our own food needs as close to home as possible. ... one part of which is having access to the seeds we need to grow food.' Ensuring seed access through seed saving contributes, in Susan's mind, to food security – not only for herself but also for her wider community. Many savers echo the importance Susan places on seed as part of regional food security.

Adam explains how strategically saving to improve seeds is integral to food security in the context of changing climates:

> I believe that seed saving, done well, can improve our crops so that they can meet future challenges like climate change, droughts, high UV exposure, etc. Seed saving can also help with important traits like taste, colour, shape that are important to the cultural side of why we eat food. As I spend more time saving seeds I have become aware that the decisions I make in the field will change the plants that grow from my seeds. Therefore, it is important to me that I make decisions that affect the plants positively and improve the crops that are grown from my seeds.

Learning about seeds and saving facilitates Adam's abilities to adapt seeds, and in making care-full decisions he attempts to 'affect the plants positively' as well as improve human food possibilities. Ada elaborates on the role seed saving plays in conserving options as well as building alternatives, saying:

> I think that the future of our food sources and the danger these sources are in has changed my ideas about seed saving. I think now more than ever before, seed saving has become a necessity in order to preserve our 'food heritage', and biodiversity. As I mentioned before, I think seed saving gives people power. The power to control what food they grow, and when, where, and how, that food is grown.

Seed saving is not only about providing food in the present, it is about preparing for future problems with food security – from lost biodiversity, changing climates or corporate control. Saving also opens possibilities for alternative food arrangements. It is, therefore, deeply entwined with making ethico-political differences.

Seed saving as food provisioning may be complementary to other ethico-political engagements such as conserving cultivated diversity, as comments at the beginning of this section from Richard and Ada indicate. Connecting food provisioning with conserving, Adam suggests:

> one of the best ways to save our, our biodiversity, is to get small-scale farmers growing it out, finding a market for it, and getting customers eating and saying, 'I really like that yellow wax bean!' and then they, that's how you really get interest in those seed varieties, you know?

Here, Adam submits that using consumers' tastes and a revised market system with small-scale growers will help conserve biodiversity. For some varieties this is the case – as we have seen, taste is a powerful motivator to keep varieties going. Yet not all varieties are tasty or appealing in some sensory way to all savers.

Maintaining less appealing varieties may be lower in some savers' priorities, while others feel compelled to save those seeds – for seeds' histories or because of biodiversity concerns. It is sometimes difficult to negotiate for oneself whether to keep or stop saving less appreciated seeds. Gail, for instance, explains that she is 'motivated to exert certain efforts on behalf of certain seeds that I actually use.' As she shows me her collection of beans and peas – stored in plastic dry-goods jugs in her extensive shed – she relays the story of one of the peas:

> So at a seed exchange there was this red, crimson pea. And red is a kind of unusual colour for a pea. And it said, 'very rare – please save seed.' I thought, 'Oh cool!' So, I took the little package of seed and planted it – grew it out. And when the time came to harvest the peas [rubs hands together, smiles] I went over and I did it and I bit into my first one and [screws up face, shudders]. Horrible. I've never grown anything like it – I don't know like a cowpea. It's just something you'd give in starvation times, you know? But on the other hand, for the sake of genetic diversity it's probably a good idea to keep that variety going. ... So that's the story that goes with this. That's almost the only time I've grown something that's [pause]. I really thought, 'Oh, I don't really think I'll grow this very often!' But I have the seed. It's collected. I should put it in a vault somewhere [laughing] and pass it on.

Gail is left with a pea she feels obligated to save again because of its rarity but that she does not find tasty and will not use. She will not grow and save it often. Perhaps by now she has passed it along. As Gail's story demonstrates, eating and conserving sensibilities sometimes clash. Not all heritage or rare varieties are as tasty as Jane's Bull's Heart or Denis's orange Mennonite tomatoes (see Chapter 6), and not all savers share the same tastes or priorities. How to deal with these disconnections is an ongoing issue in seed saving.

Seed saving joins food provisioning in the plots of many savers, and through saving, savers participate in reconstituting shared worlds. As the quoted savers illustrate, perpetuating heritage varieties, working at adapting seeds, and experimenting with new varieties is part of work undertaken by seeds and savers in seed saving to provide food that people (and others) enjoy and value within more self-sufficient, alternative food networks. In part, savers interests are of a stewardship variety – they provide sustenance for themselves and their communities, managing food in ways valued as better. Seeds, however, are not only resources in these processes; rather, seeds entice with taste, entwine with stories and constrain possible adaptation.

Conserving Biodiversity

What if one year, looking to buy seeds from catalogues, growers were confronted by absence of their favourite varieties? This was the situation that arose in the 1980s in Canada. In the context of corporate buy-outs, changed priorities, and shifts to mass markets, some favourite seeds – veggies, flowers, grains – disappeared from catalogues. As gardeners investigated, they found that not only were seed varieties going missing but genetic diversity was also disappearing. In this situation, saving became an opportunity raised by the disruption of usual growing practices; some people already saving seeds expanded their practices and/or collections with the revelation, while others, already growers, were spurred to become savers. The Heritage Seed Program, now known as Seeds of Diversity Canada, was initiated by gardeners observing these losses in 1984 to encourage growers to take up seed saving – to maintain favourites and contribute to conserving biodiversity.

This experience, motivation and collective response are evident not only in Canada but elsewhere. Renée Vellvé (1992: 17–18) explains seed saving's relevance particularly to biodiversity in the context of European communities, saying that:

> It is one thing to come to grips with the failings of our government's efforts to store our genetic wealth for future needs. But it is quite another thing to discover that, against all tides, there are indeed a number of people out there taking into their own hands the imperative to use and keep our crop heritage alive. These people are not cultivating museums, as many of their governments are. Moved by the loss of what are valuable plants, they are managing diversity in day-to-day gardening and farming: growing it, nurturing it, working with it and enjoying it. In so doing, they are preserving what are in fact our options for tomorrow:

as ordinary people, not as governments, with the responsibility to design a more sustainable food system.

Vellvé explains seed saving as a kind of situated experience of cultivating and conserving biodiversity, as distinct from practices in national genebanking and as contributions to more sustainable agrifood orders (see also Nazarea 2005; Shiva 2000). It is, he suggests, not only diversity that is being conserved in everyday practices of seed saving, but 'options for tomorrow'.

Savers' exposure to public discourse, their own experiences with variety loss and development, and their learning about diversity informs their views and practices, altering dispositions toward conserving. The term biodiversity, coined in the 1980s, pervades conservation policies, programs and discussions.[3] In this case, growers' experiences of losing favourites combine with discourses of genetic biodiversity. Seed savers articulate this combination in comments that overlap preparing for future challenges, conserving through ongoing use, and respecting biological and cultural diversity. Paul was the exception. For him, seed saving has little to do with biodiversity conservation; rather, concern about biodiversity is more about conserving wilderness. After reading savers' comments about conserving cultivated diversity in a preliminary summary, Paul elaborated his view:

> Overall I believe a far greater danger to genetic variety comes from human activity on the planet. The loss of natural habitat for vast numbers of species is like an avalanche picking up speed; it seems unstoppable. ... Surely these activities pose a much greater risk to all kinds of species and diversity; the world's gene pool must be shrinking daily. By comparison the loss of a few veggies may be harmless.

Not many disagree with Paul's sentiment that uncultivated biodiversity is important and threatened. Indeed, many savers make efforts to include consideration of this kind of diversity by, for instance, including 'indigenous' plantings, leaving spaces for natures to do their own thing, or growing seeds that are appealing to diverse pollinators. However, seed saving is primarily about cultivated diversity and most savers articulated concern for its conservation.

For almost all savers interviewed and surveyed, conserving seeds contributes, if in a small way, to global efforts of conserving biodiversity. When seed savers speak about why the conservation of biodiversity is important, the strongest tendency is to mention a future threat to human and nonhuman living due to loss of biodiversity (rather than a present loss of a favourite or some other impact).

---

3   For discussion of the term biodiversity, its meanings and its evolution see Escobar (1998, 2006); Farnham (2002); Takacs (1996); Wilson (1988). For reviews of its presence in public discourse see Anand (2006); Norse and Carlton (2003).

Bella expresses her concern about the potentially devastating impacts of biodiversity loss in this way:

> Genetic diversity is important because it allows plants and animals to weather adverse conditions such as changes in the weather, attack by other organisms and so on. ... When genetic diversity is bred out of species to promote say, uniform ripening of fruit, some other traits needed for surviving adverse conditions may be lost and the entire species wiped out. By preserving a wide group of cultivars of food plants genetic diversity is maintained.

Bella is aware of, and concerned about, loss of genetic and species diversity due to human practices. Further, the potential implications – for humans and nonhumans – make her anxious. This sense of the importance of biodiversity resounds, to greater or lesser degrees, among savers. The threat of biodiversity loss is not an immediate one; rather, it is present as a future possibility, a threat.

Although savers recognise a need to address biodiversity loss, the level of urgency is sometimes tempered with observations that seeds continually evolve. This moderation finds expression in one saver's comment in a workshop:

> The genetic diversity that we have today is not finite. ... As long as we have attentive regional farmers all over the globe that are doing what all of us intend to do, and many of us are already doing, we'll be able to latch onto enough genetic diversity that's arising spontaneously all the time to save the ship. It's just a matter of us doing that. Do we need to preserve the old genetic diversity that's already been created and that our ancestors latched onto? Absolutely, that's part of the puzzle. Why reinvent the wheel?

This complicated understanding of cultivated diversity as a continually reconstituting phenomenon (rather than a finite resource) does not, as this statement illustrates, suggest that existing diversity should be ignored. Rather, this saver indicates 'old' cultivated varieties should be conserved, while simultaneously recognising that 'new' varieties may present themselves. This view clearly resonated with workshop participants as I looked around to see numerous nods.

The abstract and future threat of lost diversity, particularly at the genetic level, may serve to legitimate exclusive expertise or foster disillusionment; however, these possibilities are not the only ones. Rather than allowing this threat to be disempowering, seed savers explain that conservation of biodiversity is one reason to integrate the practices of saving seeds into their everyday lives. Hannah remarks that when some 'crops fail due to lack of diversity as in the potato famine in Ireland, it will still be possible to have crops from saved seeds which have not failed. The remaining seeds will have necessary genetic diversity.'

In this sense, seed saving becomes partly remade by acknowledgement of threat, but also reinforces each person's role in recreating future possibilities through everyday practice. Jane puts it this way:

> Seed saving is important to me primarily because I feel that I'm doing something of value helping maintain biodiversity and protecting the widest possible gene pool, therefore protecting the future. Sounds very grand when I write it down, but by saving heirloom seed, I really feel like I'm living the 'Think globally, act locally' credo.

Jane makes it clear that she is aware of biodiversity loss and its possible implications, taking partial responsibility to do something about her concerns through saving seed. In saving seed, she feels she and the seeds are contributing to better global futures. Conrad also suggests conserved seeds offer possible future solutions in changing agri-ecosystems:

> saving some species may permit someone in the future to crossbreed some desired properties back into the species at large. I also see a possible future where conditions on earth may be different from those of today, where traditionally developed varieties may serve better than new varieties.

Conrad values the possibilities that seeds carry with them as potential solutions, noting that breeding traits back in, or even whole 'old' varieties, 'may serve better'.

Hannah, Jane and Conrad each indicate in particular ways that the loss of biodiversity threatens futures, but also that seed saving provides opportunity to do something about it. Will sums up this feeling of making a difference to conservation and shared futures as a part of seed saving: 'There is also this sense of stewarding and protecting the future, because we all know how important the genetic variability is to the world.' Will suggests that 'we all know' about the relevance of genetic variability, and he makes sense of this importance through stewarding seeds and in terms of 'protecting the future'. It is a sense not far from Berry's elaboration of agrarianism as explained earlier in that seeds are used at the same time as being maintained for future human use.

Conserving biodiversity is more than just a perk of seed saving to these savers, it is a responsibility. For some, this sense develops through practices. Lena considers this feeling as she shares her seed saving path:

> [I began saving] primarily seeds of flowers that I liked and then as I got into the seed saving more, I would look for varieties that were unusual or that were obviously heritage varieties. I would look through the listings for varieties that people said that their grandmother had grown or that they had found in a field of an old farmhouse. It seemed to me that was the purpose of the seed saving, was to

continue with the old varieties so they wouldn't die out. ... And I guess that's why I'm a seed saver – I think it's up to the individuals now to make sure that these old varieties continue.

The benefits of seed saving – for Lena initially having flowers she liked – change with experience, as do savers' senses of responsibility.

Taking action through seed saving, savers suggest, helps conserve biodiversity as part of meaningful engagements with seeds. Irene elaborates on this sense of collective contribution and connection:

> All of the reasons that so many of us save seed are about our ethics about conservation, valuing of diversity in the plant world, and valuing of the cultures that have evolved around the earth – native agricultural crops – and that resonates also with my environmental consciousness and my spirituality.

Collective benefits extend to other people in the sense of future available resources and present/future food security, but this is just part of conserving seeds as practiced ethics. As Irene suggests, there are worldly matters of concern with how more-than-human cultures connect with conservation and saving, and with general valuations of diversity in natures. In this way, she points to ways in which biocultural diversity is valued in and of itself, in addition to its usefulness to humans (present or future).

In some ways seed saving is human-centred since ensuring cultivated diversity is understood to be a good way to ensure human food security (present and future), but some savers – like Irene above – also articulate respect for diversity itself. These are not mutually exclusive positions. For example, Lena begins talking about conserving biodiversity as a risk management strategy, but ends on a different note – one of personal fascination – as she explains:

> I think that if we lose these varieties they are lost forever. The more diversity we have in the world around us and in the plant world, the better we'll be able to survive weather disasters, or ecological, or climate changes. It gives the world more options for managing change ... also it's just more interesting to have more variety.

Conrad goes further, arguing that diversity is more desirable overall: 'I think there is generally more value in diversity than in uniformity. This opinion applies to many venues of life (culture, science, politics) and also to nature.' Seed saving for Conrad, therefore, is just one way of fostering respect for diversity, something he values in many realms. Bella clarifies that conserving biodiversity is about more than managing resources wisely, asserting 'I feel we have a moral obligation to preserve the many species of plants and animals that share the earth with us.' In these statements Lena, Conrad and Bella – to differing degrees and in different ways – understand seeds, along with plants and animals, as more than resources. Bella makes it particularly explicit that she thinks we owe respect and care to the different and partially connected creatures – including seeds – 'that share the world with us.'

Becoming Collaborators with Degrees of Freedom

So far in this chapter I have focused on relations of seeds and savers in seed saving that might be considered more human-centred – food provisioning and conserving. But even with these practices, savers and seeds have made sure to complicate understanding. Seed saving intervenes to reconfigure relations among humans, seeds and their environment; but savers are not the only actors. Seeds and savers actively engage in seed saving – achieving good food and conserved seed together. The differences made matter to savers and seeds as well as food and ecologies beyond the field and beyond the present. I think on the basis of the discussion so far, it can be argued that seed saving is relational and worldly ethico-political engagement of multiple human and nonhuman participants. In this section though, leaning on Plumwood and Haraway, I push humans further from the centre of seed saving, recognising the relative agencies of seeds and savers. With this as the general aim, this section considers how savers select possibilities and adapt capacities, as well as how seeds demand accommodation and present opportunities, for those savers paying serious attention.

Savers expect and desire the minglings and complexities of active engagements with seeds. As Pat puts it:

> [s]eed saving has certainly contributed to and expanded my understanding of nature – and vice versa. ... [I was] always seeking to be out in, and connected to, nature and the natural world, and I think that connection probably opened me to the wonder and magic of seeds – and hence seed saving, sharing, and spreading around.

By acting with and in natures through saving seed, Pat, among others, learns about, connects with and opens up to seeds and natures. Pat's dispositions toward natures inform her practices, and saving practices evoke – at times – a sense of appreciation. There is also a relation of difference; Pat connects with seeds, but celebrates in wonder at their difference. Here, wonder propels further and farther engagements (see Bennett 2001, 2010); Pat saves, shares and spreads seeds – and saving – around.

But feelings of wonder are not all that come with saving, as an extract from my saving fieldnotes indicates:

Just a quick note: I do not want to save seed today. I do not want to be in the garden. I have other things to do – lecture to prep, chapter to write, papers to grade. But they are talking about frost overnight, and the seeds aren't ready. I need to cover them and hope they make it through. The plastic sheets I have will have to be gerry-rigged with stakes to hold them up and bricks to hold them down – not the best option but I hope it does the trick. Why couldn't the weather wait until the weekend? Or the seeds have been mature enough last weekend?

Not my happiest or most shining saving moment, though perhaps my questioning of weather and seed timing might generously be refigured as wonderment of a kind. With or without wonder though, I certainly felt responsibility. It was not the easiest process, and the timing was poor for the seeds and me. Not all the seeds made it through, but some did. And I was glad I made the effort, for the seeds and for my learning about weather patterns and adjusting to them and seed requirements. Seed saving keeps you on your toes, and keeps you humble – mutable nonhumans, partial knowing and ongoing processes challenge any sense of control or settled arrangement.

Savers and seeds, as earlier noted, have degrees of freedom in practicing seed saving. Seeds that self-seed, spreading themselves without savers' aid, are the clearest example of seeds acting as free agents, but most of the time – particularly with cultivated seeds – some intervention is involved. Even at the other end of the spectrum, when savers have much control such as in selection, seeds have some influence. As Haraway indicates about laboratory and food animals, nonhumans can simply refuse to survive – a low level of freedom certainly, but freedom nevertheless. Hitchings (2006), as part of his analysis of home gardening practices in a suburb of London, considers how gardeners respond to plant agency. He argues that to maximise wellbeing in gardens 'gardeners had to skilfully find a way of simultaneously taking charge and ceding control' (2006: 377). For Hitchings, this wellbeing emerges only with expert practices, while novices are left trying to figure out how to control things. This division of controlling efforts between more and less experienced practitioners does not hold for the savers I spoke with, but the idea of shifting relations of control is helpful.

Working with, rather than on, nonhumans is both joyful and frustrating, and it is through these tricky entanglements of negotiating control and working toward response-ability, that savers and seeds thrive. The understanding that savers only partially understand and influence how things turn out in saving is plain from people's discussions of their practices. This recognition, however, did not lessen the desire to engage; rather, in most cases there was challenge, joy and respect found in seed saving as partial connection with degrees of freedom.

Selecting Possibilities

An important part of saving involves selecting from those seeds that thrive – choosing which seeds will continue. These judgements matter to seeds and savers, and are almost as varied as seed-saver relations. Checking for damage and choosing healthy seed is a start, but selection is more considered. Savers examine shapes, sizes, colours and so forth (Figure 7.2). Even seeds that look the same may be differently adapted to disease or climate, or have different histories – each worth deliberation. Eating and cooking practices and aesthetic valuations come into play as growers consider the possibilities and constraints of particular seed materialities; small beans cook faster, big ones are easier to shell, and each has tastes, textures, nutrition and colours of its own. Diverse varieties, techniques, environments and genetics add complexity. In addition, savers practicing selection monitor and sort seeds' diverse materialities throughout the season and seeds' life cycles. In short, sorting and selecting seed involves material, sensate and cognitive engagements that are complicated, requiring attention to visible and invisible seed materialities.

Fig. 7.2   Separating and sorting seed

In all of this complexity, selection is done in multiple ways with different objectives – conserving, experimenting or improving. Selecting for conservation requires the greatest constraint of seed possibilities and saver activities while experimenting depends upon encouraging seed mutability, and improving lies somewhere between. There are, therefore, degrees of freedom for seeds and savers even within selection – the most saver-controlled part of saving.

Saving as conserving aims at keeping seed as similar as possible from generation to generation – this means saving 'true-to-type' seed that is 'pure' with diverse genetics – but saving with other purposes results in differentiated practices. In selecting true-to-type seed, savers watch for seeds with a 'typical' profile, asking themselves questions like: is this seed usually round or oblong?; is the flower commonly pink or white?; does the vegetable tend toward sharp or sweet taste? Seeds – and their plants – are read for difference, and major differences – rogues – are removed. This contrasts with selection practices for other purposes in which savers seek out and perpetuate seed differences – unexpected crosses, early bloomers, better reactors to stress. Selecting for experimenting requires rogues to be saved as a separate group to be explored in coming seasons, while improving aims at compromise – perhaps valuing drought resistance, for instance, but within seed generally true-to-type.

How strictly or loosely purity is pursued also varies among savers, seeds and purposes. Saving to conserve seed varieties, for instance, requires keeping seed pure, while in food provisioning purity is less important. Keeping seeds pure mostly involves altering growing practices. In some cases, impurities – genetic crosses – might be observed, for instance, in leaf forms, flower colours or seed shapes, and might be dealt with by saving true-to-type seeds or experimenting with a new cross. However, to save pure seed, savers attempt to pre-empt seed differences – whether they are observable in the field or not. Some seeds make this easy – self-pollinators like beans and lettuce take little attention to ensure purity, and clonal propagations like potatoes or garlic have no genetic impurity risks (though these materials carry other potential problems). Other seeds – like cross-pollinating carrots and squash – require savers to take further measures. Such varieties are isolated from potential crossers as much as possible using distance or time – growing in different areas or seasons. Measures like bagging and caging are other isolation options, particularly useful in smaller spaces or with promiscuous seeds. There are also implications of isolation beyond maintaining purity – growing one squash a year because it would cross with others in the family, for instance, means that there is only one to be saved, eaten and/or shared.

Professional seed growers and savers conserving seed are also those most concerned with retaining a wide genetic pool, though most experienced savers make some efforts in this regard. Too small a gene pool results in 'inbreeding depression', which shows as seeds become less vigourous, less adaptable and less productive. In order to prevent this phenomenon seeds must be saved from a variety of plants. For improving purposes, seeds are selected from several very good performers. In contrast, for conserving, seed is kept even if it is not the best performer in order to ensure genetic diversity is retained. Again, some seeds make the process easier than others – self-pollinators generally require saving from fewer plants than cross-pollinators. For those savers unable to grow as many plants as is recommended to prevent depression – which can be in the hundreds for some seeds and purposes – sharing, exchanging or even buying seed to mix in diversity is important practice.

Conserving, experimenting and improving are each part of how savers and seeds achieve seed saving. Selecting seed for conserving a particular variety – valuing typical form, genetic purity, and wide genetic pools – is at one end of a continuum of selection practices and requires relatively high levels of control. It is not for every saver, seed or circumstance. The intensity of savers' concern and their experiences, seed materialities and shared immersions combine to configure selection processes. Doing selection throughout seed lifetimes is a complex process involving multiple considerations and materialities, and savers make only some of the difference.

Adapting Capacities

As savers become sensitive to the capacities, tendencies and affects of seeds they become more able to understand how to work with, and selectively engage, seeds. This is most obvious in savers' engagements with adapting seeds, which requires a coming together of seed, saver and socio-natural affordances. Seeds as they are today are not exactly the same as they once were – whether that means a hundred years or just a season ago. Seeds have embodied changes in environments and techniques, and savers have changed with these too. As Jane suggests:

> Each locality has its own set of conditions that affect growing – soil types and weather, number of growing days. Not all varieties will do well in a given location. Varieties that do well locally should be maintained. These are not necessarily the varieties produced by the large seed companies.

The different practices at larger seed companies and multinationals result in different seeds, perhaps maladapted to local conditions since they have been bred to survive – though perhaps not thrive – in several climates. Moreover, growing spaces are expected to adapt to seeds with chemical and mechanistic methods. Saved seeds, on the other hand, adapt capacities and alter dispositions over years spent in places with savers, learning and adjusting to particular climates and techniques.

Interest in locally adapted seed is a large part of why savers practice selection and saving. Not all seeds will grow all places, but those that do grow adapt over time, providing new opportunities. Richard makes his interest in adaptive seeds clear:

> The farm is an ecosystem, and the seeds that are grown there have to be integrated in this ecosystem. We cannot rely indefinitely on seeds that have no local adaptation and don't have the built-in genetic diversity who permits the plant population to adapt to different seasons, soil variations, different cultural methods.

Likewise, a surveyed saver wrote:

> When I collect and store seed obtained on my own farm, I know the history of the
> parent plants and their performance. Although I am still a novice seed collector, I
> have confidence that slowly, over time, I will create a diversity of plants especially
> suited for my farm's growing conditions.

For Richard, the capacities of seeds to adjust to differing contexts are necessary.
The surveyed saver, however, emphasises her own role in enacting these changes
with seeds and environments over the long-term, though the seeds' performance
on-farm is noted along with the need to take time in adapting seeds. Each
statement, despite different emphasis, demonstrates simultaneous relations of
control and reliance among seeds and savers.

Savers recognise relative agency, nudging humans from the centre of saving
practices and insisting on agential seeds (among other nonhumans). Sandra's
guidance for new savers is illustrative, she instructs:

> Get to know the plant and what is requested for next season['s] growth. Wait and
> watch Mother Nature and pay attention to the weather. Leave the seed on the
> plant for as long as possible. Have a place where you can dry the seed. If possible,
> ask questions [about] what others have experienced. Label the bag with the right
> seed in it.

Here Sandra makes it clear that savers must: figure out what seeds request for
growing; wait and observe nature's processes; accommodate growth cycles;
provide space and dry conditions for storage; learn from other savers experiences;
and, take care in organising. A majority of these instructions place seeds rather
than savers in the middle of practice. Gail offers another way of thinking through
seed agency, explaining:

> Plants over the millennia have adapted themselves to local environmental
> conditions. Farmers traditionally have studied their plants and saved the seeds of
> plants most adjusted, to their local conditions. In my garden I continue the same
> tradition of saving seed from my 'best' plants adapted to my local conditions.

Here it is plants doing the work of adapting, with growers taking opportunities
to collect adapted seeds to get better seed – and plants. Accommodating seeds, by
learning and adjusting to seed capacities and timings (as discussed in Chapter 6)
and learning to recognise and take up possibilities presented by seeds, is one of
the ways it becomes so clear to savers that they are not in total control and that
seeds are not the only organisms changed through saving.

Demanding Accommodation

Successful seed saving requires accommodation of the living natures of seeds. Some measures – like isolating, fermenting, second year growing and protecting from frost – have already been mentioned. In addition, plants-becoming-seeds require adjustments in growing spaces, and sometimes this means eliminating other plants or supporting seed-plants with stakes or trellises. Some seeds need scarification, soaking or stratification before they will germinate. In a conversation about grains and threshing, Phil offers insight into how some seed bodies constrain savers. As we are walking his farm, he informs me:

> I have a threshing box that I use with my feet – I thresh just rubbing my feet against a few slats on the bottom of the box [demonstrated gesture with feet]. And almost all these – the ones that thresh easily – thresh easily. And the other ones, there's nothing you can do. [shaking head] They might have possibilities for farmers on the prairies, but, unless you buy a small-scale threshing machine or something. But we're not into that yet in North America. It's very difficult to find machines that will thresh small amounts, like home garden amounts. But it's very common in Third World countries to have threshing machines powered by bicycles or hand-cranked. And they can be quite efficient.

Technique has much to contribute here in dealing with seed materialities, but with or without particular technologies, seeds must be accommodated. With some grains, in his situation, 'there's nothing you can do', so Phil concentrates instead on hull-less or easily threshed varieties, doing bulk threshing with feet and wooden threshing box.

Different seeds therefore require different engagements. In some cases, a change in variety or technique is necessary. In other cases, one admits defeat or just learns seeds better. Squash, for instance, crosses easily, which proves challenging to those unaware of the seed-plant practices. Savers told many stories of interesting (sometimes bizarre) crosses they came across while growing squash. Brent reveals that his saving squash both required attention and provided opportunity for him to learn more about pollination processes and methods. He now hand-pollinates squash to ensure quality seed, and he explains: 'I had to pay real attention to that [hand-pollination]. It takes knowing male and female flowers and knowing when the flowers actually open – they open in the morning, so you have to plan it.' He has worked to accommodate squash not only by learning the seed-plant's needs and habits, but adopting new methods and timing. As Alice notes, '[s]eed saving is a challenge. It's always a chance they won't germinate. ... The more dubious I am about the seed, the more exciting to see the first sprouts.'

There are plenty of methods that improve chances with seeds, but in some cases, savers just have to wait and see. Adam reveals this as he shows me a particular onion crop that he is quite excited about:

> It's a gorgeous crop. As you can see it's doing very well, it's really healthy. And I'm just waiting. So I'm watching all those bees in there pollinating and I'm just waiting for, I hope everything goes good and we'll get a good seed crop.

Adam does all he can, and watches attentively, but sometimes he must just 'hope everything goes good' and good seeds present themselves. Making savers wait as they reproduce, grow and mature is another way that seeds demand accommodation. Seeds provide challenge and constraint, and also offer opportunities.

Presenting Opportunities

Seed capacities and tendencies can constrain saving possibilities in many ways, but seeds also present possibilities. In other stories of processing seed, savers explain that walking or stepping on dried bean and pea pods to break them open before sorting is faster – and much more fun – than breaking each pod open individually given the bulk amounts some savers deal with, and the strength of the dried seeds makes this possible. Sometimes other savers help in this activity (see Figure 7.3).

Fig. 7.3   Crushing pods and selecting beans
Source: Wildfong 2012

In learning seed saving one becomes more able to discern seeds and their changing needs, as well as the practices that might work with these changes to achieve saved seed well, but sometimes the seeds make it so easy and obvious that to imply that they are doing anything less than inviting or even suggesting saving practices seems inadequate.

Many savers relaying their first experiences of saving seed explain that it was the surprise of seeds showing themselves that evoked thoughts of saving. Ada, for instance, indicates that she began saving 'purely by accident', further detailing, 'I got busy one summer when I had a community garden plot. ... and before I knew it my plot was full of mustard greens, arugula, and marigold seeds!' This is not an uncommon story, and the event of seeds surprising *growers* can be frustrating in the sense of lost plants, but to *savers* the possibilities invite response. Natalie describes getting started in a similar way:

> The first crop I have ever saved seed from in my own garden was Red Oakleaf Lettuce. It went to seed in the heat of summer and I decided to let it. The success was so high that we allowed Champion Radish to go to seed as well. After that I was hooked.

Natalie and her husband now save a collection of seeds, sharing them widely in their community. But it was lettuce seed appearing through bolting plants that encouraged her. Further, it was the success of that surprise encounter that got her 'hooked'.

The capacity of seeds to reproduce themselves – sometimes in astounding numbers – also provides possibilities for sharing, experimenting and eating. Brent reveals that he thinks of seed saving as 'a great investment ... with one bean seed you get – I get 25–50 but I'm sure you could get more than that. There's nowhere where you can invest a dollar and get $25, you know? I always think that's good.' For Brent, the reproductive capacity of seeds provides an excellent return considering their requirements. In a similar vein, Phil shares an example of amaranth, saying:

> but these plants [gesturing to amaranth] are amazing because they grow to be usually 8 or even 10 feet high. And they produce so much seed, it's unbelievable. One plant can produce a quarter of a million seeds! ... They start off as just a tiny little seed, smaller than a millet seed.

When the amaranth seed is ready, Phil collects it by carefully bending the plants over a bucket or wheelbarrow with a screen on top and then shaking or rubbing the flower-seed heads so the seeds fall. Wind carries the chaff away, or it is later blown out. The process is much like that for lettuce, though with a much larger plant and one likely to still be flowering. Amaranth seed is highly nutritious and can be cooked into porridge or popped. It tastes slightly sweet and a bit nutty. Phil explains he loves to eat it and share it with others. The fact that amaranth produces so many seeds opens these options – if there were fewer seeds more plants or more restricted use would be necessary. There's also a sense of amazement in Phil's story of a tiny seed growing into such a large and beautiful plant.

Wonder is part of the appeal of amaranth, but other seed-plants – less extreme in their material forms – evoke similar responses. Dara explains that she enjoys sharing foods with people, and she is particularly fond of arugula. She elaborates: 'When you are saving your own seed, you tend to get a whole bunch of something. When saving arugula I end up with a huge bag of it. I give some away, and it gives me the opportunity to experiment with different growing techniques.' So Dara gets to try new things for growing, as well as sharing produce with others because of seed saving. Rachel also notes that seed facilitates sharing, but she tries to share her flower seeds because there is simply too much for her to grow herself:

> I'll have so much seed I won't know what to do with it. I couldn't possibly plant it all. With poppies, or columbine, or calendula, or whatever, there's just so much seed that I couldn't possibly use it all. And so I try to give it away. But I think sometimes people don't know what to do with it, or, they're not sure. People, I think, are sometimes afraid to try. And I'm not.

Rachel maintains that seeds may present opportunities, but only some people take action on the chances. As she tries to share her seeds, she encounters people who 'don't know what to do with it' and are 'afraid to try'. But she, along with other savers, tries things.

Experimenting and seeing how to help seeds thrive is part of the challenge and enjoyment of seed saving, and in these cases there is plenty of seed to try with. Monica, particularly interested in alpines and rock garden plants, elaborates '[s]ome are difficult to grow successfully and purchased plants are often quite expensive and I can't afford to throw that money away so growing from seed gives me a better chance to get plants growing that will tolerate my conditions.' Here seeds provide opportunity to save money, and to experiment and learn with more challenging plants within a particular place. Not all seeds thrive here, but some do and they continue in reconstituted lives with Monica and those with whom she shares.

Collaborating with Seeds

It is clear that seed savers are not the only beings acting in growing places. Further, though this lack of control can be frustrating or confounding at times, on the whole seed savers enjoy that seeds do their own thing. As Lena explains, 'The seeds can grow by themselves, but they would depend on me for soil and water and the right conditions. It is rewarding to plant a seed and watch it grow and blossom and do what it's supposed to do.' Here seeds have their own abilities and purposes that Lena simply tries to support. Further, she finds this supportive relation rewarding. Rachel sharpens this sense of enjoyment of seeds as living creatures with whom savers partially connect as she reveals:

> I don't think that I understand all of it. I don't understand all of it. But I have some respect that it lived before me, and I hope that it lives on after me ... I just want to let it do what it does and witness that, and be present for that and just appreciate what it is. And not get in its way [laugh]. ... I like that it grows and it changes without me. I REALLY like that.

Part of the pleasure of saving, for Rachel and others, comes with attachments to seeds that are more than resources. Understanding and appreciating seeds as creatures with their own lives, capacities and dispositions is integral to saving practices.

Seeds become co-actants and participants in seed saving. Not all, but many savers, practice a kind of selective anthropomorphism, talking about how seeds 'are really liking this' or 'aren't happy there'. This tendency will not surprise those people who know, or are, growers and have heard many tales of seeds or plants doing and relating in ways often reserved for humans, or sometimes pets. Seeds are more than inanimate matter, and savers try to explain their relations in various ways. This is not a managerial or ownership relation, or at least it is not only that. Beth explains her own view saying, 'They're not *my* plants, I'm just taking care of these things – like the trees in my yard, everything. I'm just here to take care of them, they're not mine. It's *not* an *ownership* thing.' Alice, repeating other savers' comments, contends: 'when you have grown the parent plant and then collected the fruit (seeds) it's like you have a partnership.' In some cases seeds are partners, and in others they are friends or companions. They may also be likened to children, as in Will's remark: 'Our chestnuts and walnuts and other nut trees are connected to us in a very similar way as parents feel about children. We take care in planting, protecting and fertilizing just a little more.' Plenty of options beyond stewardship of resources articulated themselves for describing the complex ethico-political relations among seeds and their savers.

It is not that seeds are the same as people; rather, savers attempt to find ways of conveying how they relate with seeds as active agents in seed saving – as differentiated beings, in unequal and ongoing partnerships. Brent sums up the connections of seed saving as a continuing collaboration with seeds by saying:

> I feel the connection with nature through seed saving because I've seen a plant from seed to seed. I am connecting with the plant from the moment a seed is put into the ground. I co-create the environment in which it grows. I think of us as

in it together, with the plant doing most of the work and me clearing the way for it to flourish. In the fall, when the seeds are ripe, I harvest them, dry them, and store them. In a sense I am the caretaker of the plant until the next spring when its promise can be achieved once again.

Brent suggests here that he and the seeds are 'in it together', and this idea was prevalent among savers. The connection with the seed's living – 'from seed to seed' – is facilitated by Brent, not as a sole controller but as a co-creator of seeds and their socio-natural worlds. He clears 'a way for it to flourish' and achieve its purposes by participating in careful and responsible seed saving.

In sum then, seeds and savers practice saving – in all its material, sensorial and cognitive complexity – together. Seeds are, like other nonhumans, understood as 'partially connected to humans: they are partly being made together. But their specificities are not human. Their relations also extend, and are performed, elsewhere' (Law 2004: 5). By tying their future possibilities with seed saving practices, seeds gain certain privileges (like regular tending) and exert certain controls (like refusing threshing). Seeds invite through surprise, enchant with story, frustrate with timing, and demand accommodation. They do their own thing – self-seeding and making savers wait, act and hope. In all of these ways, among others, seeds actively participate in what Haraway calls companion relations. Seeds open up and shut down possibilities for themselves and others. For their part, savers learn seeds and techniques, adjust to and adapt seed capacities and tendencies, recognise and seize opportunities, foster flourishing, and selectively kill and eat seeds. This collaborative (or companionable) relating is fundamental to seed saving as successful and continuing practice, first, because seed savers are interested in complex entanglements with seeds, and second, because it is only through seeds thriving that saving is possible. In seed saving, mutual flourishing – not of all, but of enough to continue – is necessary.

I have argued here for an understanding of seed saving as ethico-political practice – worldly engagements among seeds and savers that make differences to collective futures. I have suggested that seed saving includes human-centred activities focused on present and future needs, such as food provisioning and conserving biodiversity, but that even these pursuit become more than just careful resource management when closely attended. The ethico-political praxis of saving in this chapter oriented around intimate and collaborative engagements, with some attention to making differences in wider worlds. Recalling Haraway's mandate to make ethico-political relations unavoidable in everyday living, this conversation continues in the next chapter, which explores the politics of saving in ways focused on primarily human socio-economic relations.

Chapter 8
Resisting, Remaking and More

> Remaking the food system then suggests neither a revolutionary break nor a radical transformation but rather deliberate, sometimes unglamorous multipronged efforts in areas where openings exist to do things differently. ... Such activities quietly and modestly remake parts of the food system. Whether pursued by individuals, by groups, or by communities, such remaking is not a linear or foreordained process that possesses some clear, known endpoint. It is instead movement in what is hoped to be a more promising direction. Remaking shifts us from a paralyzing focus on what is worrying, wrong, destructive, and oppressive about our current food system to a wide-angle view that takes in the broader landscape, whose troubling contours, we begin to notice, are punctuated by encouraging signs of change.
>
> C. Clare Hinrichs, Remaking the North American Food System

Seed saving participates in more accepted ways of being political in our society; however, it is also, and more often, a direct, local, embodied engagement through which savers save and share seeds in their everyday lives. Like the food-related movements referenced by Hinrichs (2007) above, seed saving may not be a 'revolutionary break nor a radical transformation', but it is a way of continuing historically useful and valuable practices, and of pursuing and initiating openings 'to do things differently'. Saving works with other efforts to change agrifood arrangements in seemingly small but vital ways. In an examination of opportunities for economic change, Gibson-Graham (2008) cite a myriad of ongoing projects – several reflecting Hinrich's idea of remaking food systems. They suggest that rather than reading for revolution or resistance, analysts should re-read for difference and cultivation of creativity. In all their diversity and constraint, such projects are, Gibson-Graham suggest, ongoing attempts to change the world for the better, rather than inadequate revolutions against capitalism.

This chapter furthers my argument that seed saving is ethico-political engagement. The previous two chapters considered seed-saver relations in relatively intimate terms, with Chapter 7 offering some connections with broader issues of food production and biodiversity conservation. This chapter, keeping Hinrichs and Gibson-Graham in mind, continues developing the idea that in saving seed people work with seeds to remake, or recreate, alternative seed (and food) possibilities. However, complementing this kind of rereading, seed saving here is also considered as resistance. The following sections address savers' efforts as: resisting corporate control in the face of governmental shortcomings, including responding to genetic engineering and intellectual property rights; contesting commodification; re/creating self-reliance; fostering collective belonging; and cultivating alternatives. Before exploring saving in such ways though, I consider political engagement more broadly, first connecting savers with more official politics and then outlining a more everyday understanding.

Political Engagements

Isin (2002: 275) proposes that '[b]ecoming political is that moment when the naturalness of the dominant virtues is called into question and their arbitrariness revealed.' For Canadians, several such seed-related moments have presented themselves recently – introductions of and contaminations by GE crops, proposed policy moves restricting seed saving and promoting GE, advocacy by industry and government for stronger property rights on seeds, cuts to public plant breeding programs, dismantling of the Canadian Wheat Board, and Canadian defence of Terminator seeds. In each instance, the values underlying corporate seed ordering (such as genetic and economic reductionism, private property rights, and instrumentalism) emerge as contestable and contested. While the best use has not been made of these opportunities for changing agrifood or seed relations, savers have 'become political' in official and everyday ways.

Official policy systems in Canada allow few opportunities for public debate about seed (and food), and perhaps even fewer for initiating policy changes based on such debates (see Abergel and Barrett 2002; MacRae 1999; NFU 2012b). Further, any sway growers had in such systems historically has eroded as urban and corporate influence has grown (Winson 1993). Franklin (1999: 121), in considering democratic governance, argues that the institutions of Canadian government are no longer responsible or accountable to the people, but instead these institutions serve to ensure that Canada is 'safe for technology' and, one might add, its profiteers. As suggested in previous chapters, finding gaps in which to contest and alter these agendas is challenging, but not impossible.

Seed savers, along with other food activists, make demands individually and collectively upon governmental institutions to facilitate and incorporate into policy-making public discussions about food and seed issues of common concern. Letters and phone calls to political representatives, petitions and public events organised by individuals and non-governmental organisations are ways in which seed-related advocacy has occurred. Initiatives of this kind have, for instance, demonstrated public opposition to the commercialisation of GE wheat and to proposed PBR Act changes limiting seed saving and selling. In each case, public opposition played a meaningful role in stopping (at least temporarily) the coordinated moves of seed industry and government to neoliberalise seed and its relations.

In addition to reactions to particular public issues, seed savers and their supporters engage in official politics through participation in organised associations. Several civil society organisations, with saver members and seed-related mandates, serve as exemplars. Seeds of Diversity Canada (SoDC) is most commonly mentioned by savers as an organisation to which they belong. Part of the appeal of SoDC is the information it provides to savers and its facilitated seed exchange – approximately 3,400 varieties were accessible among members in 2011. Recently it has also joined with USC Canada on a project to develop seed networks and training across Canada. In addition to fostering alternative seed access and saving, SoDC representatives participate in policy discussions, especially around biodiversity, and contribute to national genebanking programs – doing grow outs and research, as well as banking varieties not already part of the genebank's accessions.

Several other non-governmental organisations also spend part of their time engaging in official politics. The Canadian Organic Growers, among other operations, has served as a researcher and advisor to the Canadian government on issues relating to organics – including seeds. The National Farmers' Union of Canada (NFU) – politically active on a wide range of agrifood production issues – recently rejuvenated its Seed Campaign begun in 2004 to resist stronger seed IPR, lobby for better resourcing for public breeding, and insist on the value of seed saving practices. Research and organising by the NFU have been key to seed-related campaigns such as those resisting GE wheat commercialisation, expansion of certified seed markets in the case of GE flax, and UPOV91 implementation (see Chapters 3 and 4). As a member of La Via Campesina, the NFU also brings these concerns to international arenas. Finally, Food Secure Canada participates in agrifood policy debates generally. FSC's food policy, based on grassroots workshopping, includes policies supporting food sovereignty principles and farmers' rights, as well as cautions about GE and corporate seeds. The list goes on, and continues growing.

Seed saving enrols and is enrolled within official politics – as farmers' rights, biodiversity conservation, and as part of food sovereignty. Seeds and growers do, as outlined, become part of official politics in diverse ways – consulting, researching, negotiating, challenging, and supporting particular seed-people ordering. However, as Kuyek (2004b) suggests, public dissatisfaction with the existing Canadian agrifood order garners limited official, organised resistance. Cooperation among producers is relatively limited, as is that among consumers and the connections between producers' and consumers' concerns and actions (e.g., on GE foods or seed saving) remain minimal. Policy agendas and discussions in Canada remain expert-driven and -oriented. Efforts to reconfigure food relations to be more sustainable, just, and/or democratic manifest in local communities and global networks of solidarity, rather than transformation of national or international ordering (Kuyek 2004b). But seed saving is not confined to official politics or organised revolution; seed saving also does ethics and politics otherwise.

Everyday Political Engagements

Resistance, once understood as organised opposition to domination by institutionalised authorities, has been reconceived, along with power, as more pervasive and relational, connecting official and ordinary political practices (see de Certeau 1984; Foucault 1990; Rose 1993; Scott 1985). Everyday practices, as they enact different worlds, affect other political relations – including official policy-making – whether or not people intend it to do so (Kerkvliet 2009). Further, as Kaplan and Ross (1987: 3) note: '[t]he political ... is hidden in the everyday, exactly where it is most obvious: in the contradictions of lived experience, in the most banal and repetitive gestures of everyday life.' Reconstituting worlds, then, begins with mundane relations.

The politics of everyday living are perhaps less visible than other, more commonly accepted means of political engagement such as protests or even letter-writing campaigns, but are vital none-the-less. Reflecting this sentiment, Vellvé (1992: 17–18) indicates that despite their significance seed saving practices go 'unseen, unrecognised, unsupported and unvalued. Picking up the work where the official sector is going wrong, they are the hidden but fundamental cornerstone of securing a better future for agriculture, one which doesn't just exploit but maintains and rebuilds constantly.' In this view, seed saving collaborations become part of politics as savers and seeds work together in attempts to improve their shared worlds.

Everyday practices, however, are not straightforward. One set of practices may manifest and represent in multiple ways, connecting and disconnecting with other practices and orders. Giard (1998) provides an example of how everyday practices can combine in complicated, sometimes contradictory ways, through a study of women's everyday cooking practices. She maintains that 'alimentary habits constitute a domain where tradition and innovation matter equally, where past and present are mixed to serve the needs of the hour, to furnish the joy of the moment, and to suit the circumstance' (1998: 151). With a caveat that cooking is not always this to all women, Giard's explanation of cooking as traditional and innovative, useful and joyful, past and present might be extended to seed saving. In my own saving, for instance, I conserve but also eliminate varieties, compost and use grow lights, celebrate and contradict tradition, and so on. Seed saving, like other everyday practices, does not exist in only one place, affect only one world, or demonstrate only one kind of relation.

The resistances everyday engagements offer occur in moments and spaces of possibility (Lefebvre 2002) that provide opportunities to become political. These moments are not only exploited by saving, but in some cases created or pried open by it. Christie (2005: 305–6) suggests that growing practices, which I extend to seed saving, 'are not always explicitly political' but that through such practices 'arise conflicts that render visible the politics of everyday life.' In taking advantage of moments of possibility, saving practices may 'elude discipline without being outside the field in which it is exercised' (de Certeau 1984: 172). These complex lived relations relate to governmental and corporatist strategies most often in a 'friction of "rubbing along" rather than in direct conflict' (Highmore 2002: 159). In this way, everyday practices – including seed saving – are not always resistant but that possibility often presents itself. Saving depends upon and reinforces certain realities while challenging others, at particular junctures of practice, time and space.

The effectiveness of everyday politics in changing dominant orderings is disputed; it seems unlikely to some that ordinary activities would prove so consequential. Such concern is expressed, for instance, in scholarship regarding alternative food networks in which, on the one hand, loose networking of many initiatives is argued to be a source of vitality, diversity and strength, but on the other hand, it is claimed that to create real change a more unified movement is necessary (*cf.* Hassanein 2003; Henderson 2000; Magdoff et al. 2000). At issue is how and to what degree everyday practices – growing food, buying coffee, home-cooking meals, participating in community-shared agriculture, saving seed – work toward 'remaking the food system' in more just and sustainable ways. Further, whether convergence across such diverse efforts is desirable, or even possible, as well as how to achieve such collective resistance is an ongoing debate (Allen 2012; Hinrichs 2012). Caught up within these debates is the role of everyday practice as ethico-political engagement that matters.

Although everyday practices may seem insignificant in reconfiguring dominant orders, practices – as lived, repeated enactments – are how orders are maintained and remade. Scott (1985: 192) reminds us that 'the aggregation of thousands upon thousands of such "petty" acts of resistance have dramatic economic and political effects.' In much of his work Scott (1985) argues that resistance is not necessarily open or official, instead domination is challenged along a range of activities including collective rebellion as well as hidden acts of diversion. Nazarea (2005: 158) points clearly to seed saving as an everyday enacted politics of this kind, arguing that:

> Although there may be a strong tendency to dismiss seedsavers' individual acts of nonconformity as idiosyncratic (which they are) or mere noise (which they are not), the fact is that they are almost universally present in every agroecological zone and cropping system and their presence and practices 'add up'. Ubiquitous and irrepressible, they perpetuate disorder by getting away with not planting what is popular, profitable, prosaic, or prescribed.

Nazarea's view of savers is inspired by de Certeau's (1984) ideas of tactics – everyday acts that momentarily subvert controlling orders. She reminds us that though saving may not always be obvious, savers are 'ubiquitous and irrepressible' and persist in their practices. As she suggests, savers do 'perpetuate disorder'. Moreover, they combine this disruption with re/creation of alternatives.

Concern about the possible effectiveness of everyday politics is also articulated in the idea that these networks may actually serve to support, rather than challenge, corporate and neoliberal ordering. Drawing on alternative food network literature again, some analyses caution that such efforts may not question the values and processes upon which dominating networks rely, but, instead, reinforce them (see Allen and Guthman 2006; DuPuis and Goodman 2005; Guthman 2007). Buying organic lettuce from a supermarket for example, may provide a consumer with organic salad and support limited use of chemicals not approved as organic, yet the purchase does little to confront individualisation, privatisation, consumerism or productivism, corporate control, and so on.

However, not all efforts need to be revolutionary, and few will meet the criteria of perfect resistance. There are many different ways in which alternative food networks – or other such efforts – can (fail to) change existing relations and worlds. Recalling Hinrichs and Gibson-Graham, such projects may be read otherwise – as cultivating creativity and difference instead of failing to replace corporate and neoliberal agrifood. Food practices, alternative or otherwise, like all everyday activities, are not necessarily demonstrations of resistance or creativity, though they may be. The point, then, is to continue to engage and examine emerging alternative networks as 'mutually constitutive, imperfect, political process ... [that] make each other on an everyday basis' (DuPuis and Goodman 2005: 269).

Having these sympathetic critiques in mind, the articulated intents of those involved may also be considered. While many activities, including resistive ones, may fail to achieve their intended result (or those of analysts), 'however partial or even mistaken the experienced reality of the human agents, it is that experienced reality that provides the basis for their understanding and action' (Scott 1985: 46). Further, it is through everyday practices that people intend to change seed-people relations. This chapter furthers explorations elsewhere in this book about why people engage in the ways they do, and what they mean to accomplish in doing so. It is rarely the case that practiced relations are either convivial or conflictual, resistive or creative. Instead, practices and the worlds they form interact in diverse ways, demonstrating compliance, support, ambivalence, indifference, resistance and so on. As such, in considering seed saving and savers' response-abilities, I look not for resistance or for creativity, but for both of these, and more besides.

Saving as Political Engagement

Saving seed always has political implications and affects, but some seed savers are more explicit about their practices as political engagements. These savers, at least part of the time, intend their seed saving to reconfigure power relations and to convey their views of existing systems within which they live. These savers believe and act as though seeds, and saving them, are of collective concern. Adam, for instance, suggests this intent as he describes how he came to begin saving seeds, declaring:

> I first started it [seed saving] out of a concern about concentration of ownership in the seed supply, globally and locally. It was important to me to be part of a movement of local seed production for local use. Growing organically, and growing open-pollinated heirloom varieties was part of that interest and concern. I believe in an agriculture that is based on local production for local consumption.

For Adam, resistance to privatising, concentrating corporate control of seeds drove his interest in seed growing. He explains his seed saving as resistance and as participation in a local movement to rebuild local and organic seed and food production. Similar motivation informs Ada's connection of saving with power and some of her broader concerns as she explains:

> I think now more than ever before, seed saving has become a necessity in order to preserve our 'food heritage', and biodiversity. ... I think seed saving gives people power. The power to control what food they grow, and when, where, and how, that food is grown. Seed saving and food growing can be done by all. It knows no boundaries. It doesn't matter if you are fat or thin, male or female, rich or poor. Anyone, regardless of race, religion or creed can hold the enormous life that a seemingly small seed possesses, and from that same tiny seed help to create new life with little to no resources needed, that is pretty powerful stuff.

Ada believes that saving seed 'gives people power' including control over their food choices, and that it contributes to conserving biocultural diversity. Further, she combines this with wonder at the power of seeds themselves, and a sense that anyone can save seed. Of course, people's abilities to save seeds are limited – by laws, resources, knowledge. But for Ada seed saving is an accessible way of enacting a better world in ways both material and magical.

How seed savers like Adam and Ada reflect on saving practices as political engagement orients the remainder of the chapter. Every interviewee, and many savers besides, saw saving seed as a way to make the world more livable – for themselves and others. Savers place particular emphasis on resisting corporate control, contesting commodification, re/creating individual and community self-reliance, fostering collective belonging, and cultivating alternatives. Each of these subthemes of seed saving as everyday politics is explored in turn.

Resisting Corporate Seed Ordering

For many savers, state facilitation of economic and technological rationality that favour large corporations, though not unexpected, is a source of frustration. Paul, for instance, comments that 'our governments appear to be sympathetic to big business needs, very often they do not coincide with what might be beneficial to the public at large.' Others, like Phil, make it clear that they distrust government objectives and policy-making:

> I don't trust what our government's going to do until we get a real change in the way we do politics and how money oriented it is, and how you can't really do anything unless you're in the hands of, in the pocketbooks, of these people with the big bucks.

These savers, among others, feel that public policy and governance has been overtaken by economic imperatives as governments work too closely with 'these people with the big bucks.'

Belief that governmental institutions are not able, or willing, to act in the public interest is indicated by many seed savers. In reference to the government's support of GE and the development of biotechnology in Canada, for instance, Alice reflects: 'We already have all the technology we need, without genetic engineering, to feed everybody on earth wholesome, health giving food. What we lack is the will to do it. That's not a scientific problem, is it?' For Alice, governments wrongly frame and address world hunger as they promote GE as the only solution. She further implies concern about the language of science being used to veil what is a political failure. Whether savers fatalistically accept government's cooption or whether they simply lack trust in authorities' judgements and effectiveness, savers find ways of expressing their concerns that may include, but also exceed, official politics. Lack of governmental action in the public interest pushes some people to take things into their own hands – by saving seeds.

Seed saving resists particular socio-economic relations of power, especially targeting corporate control of seed networks. Savers express grave concern and sometimes outrage about increasing corporate control of seeds. Surveyed savers 'strongly agree' that they are concerned about corporate monopolies in the seed industry; at over 84 per cent of respondents expressing strong agreement, this represents the strongest response to over 12 statements provided dealing with various seed-related issues (climate change, growing practices, genebanking, etc.). Moreover, evading corporate control by supplying their own seeds was cited by those surveyed as one of the most important reasons to save seed, with offered responses such as: '[to] protest and protection from Monsanto and similar corporations'; 'independence from corporate control'; and 'maintaining seed control (because large corporations do not share our vision).'

For some, opposing corporate control was part of why they began to save seeds, while for others techno-political awareness has grown with their immersion in seed saving. Anna explains how her commitment to saving seed increased as she became more knowledgeable about agrifood: 'I began to read further into the politics of organic, and began to realise how fast the biotech industry was trying to infiltrate our food supply. I was outraged and alarmed, and wanted to do something to counterbalance the assault of for-profit mentality.' By saving seed Anna argues that she is subverting corporate control, as well as the underlying valuation of seeds as commodity-technologies. Hannah articulates her general anti-corporate feelings, and contrasting support for seed saving, this way: 'I dislike large corporations which invariably concentrate the bulk of all wealth in the hands of the greediest and most dishonourable humans on earth. I like the independent feeling of knowing that they don't control, and therefore own, the very food I eat.' Hannah's feeling of independence from corporations and accompanying sense of increased security reiterates articulations of many seed savers.

Seed saving may be relatively rare or considered only a minor act, but seed savers believe it challenges corporate control and valuations of seeds. As one surveyed saver explains, one of their most important reasons to save seed is that it 'makes a small effort to defeat or at least somewhat frustrate the corporations that want to control seeds.' Melanie suggests that the rarity of the practice and its sensibility is exactly the point, saying:

> I think seed saving is little practiced. I rarely (one in ten maybe) meet anyone who really grows anything, let alone cares if it goes to seed. I think an awful lot of the world is working on World Bank time and something as realistic as seed saving is counter-capitalist society.

For her, seed saving offers a different reality. Resistance, then, is perhaps less obvious in ordinary practices such as seed saving, but is still relevant, even vital, to savers interested in evading control and limiting profit-making by corporations.

For many, corporate control brought with it other issues of concern – genetic engineering, patenting, unsustainable production and food insecurity. Jane expresses these widely held connections while speaking about how her seed saving pushed her to think of broader issues and distant communities:

> The more I researched, the more incensed I became about genetic modification of food seeds, including the insertion of genes from other species; and the fact that companies like Monsanto are developing seeds that cannot be saved – therefore

forcing Third World farmers to have to buy their seeds each year, rather than being able to save them from year to year. Don't get me started on Monsanto – they also create genetically modified seeds that produce plants that are resistant to Monsanto's chemical herbicides, thereby creating a growing market for their chemical products as well as their seeds.

Jane's concerns about the corporate control of seed are not isolated to the impacts of that control on her own life but reach beyond her practices to concerns for other growers and other species. Gail puts it another way: 'Given that the world's supply of seed is rapidly falling into the hands of fewer and fewer multinational companies in recent years, I think that the issue of global seed saving is becoming of vital importance to the security of the world's food supply.' For Jane and Gail, therefore, relations with seeds are part of wider worlds and concerns. Seed saving emphasises consideration of ethico-political relations, and provides opportunity to change lived realities.

Seed saving practices overall are understood to be part of relations other than those serving corporate interests. Andrew elaborates this difference arguing that seed saving 'should be contrasted to seed patenting or other kinds of commercial strategies for controlling the availability of seeds. ... [Seed saving] is about conserving and most commercial strategies have to do with limiting access and converting it to a selling transaction instead of a sharing activity.' Seed saving then, offers a different way of living with seeds in comparison with a corporate ordering. Brent reiterates earlier points about resistance to corporate seed-related practices:

> The control of our food system by corporations such as Monsanto leads me to want to have nothing to do with them. That means saving my own seed or supporting smaller companies that grow their own seed. It's hard for me to say how much of my seed saving interest comes from wanting to avoid corporate agriculture as opposed to my awareness of corporate control coming from my interest in seed saving.

Here resistance becomes so entangled with seed saving that separating one from the other, and determining which fosters the other, is too difficult.

Intellectual Property Rights

In practical terms, one of the ways in which corporations control seed is through obtaining intellectual property rights (IPR) – patents or plant breeders' rights. For savers resisting corporate control, IPR present a problem and overall savers suggest significant changes to Canada's legal framework in this regard (see also Chapter 4). In the survey, for instance, 65 per cent of savers indicate strong disagreement that patents benefit growers. Most commonly savers offer brief and clear statements such as 'No patents on seeds', but some elaborate with comments like 'No living thing should be patentable. No being should own any other being or any part of any being.' Perhaps one of the more evocative statements comes from Melanie, who states: 'Seeds just should not be patentable because it's great for a few lucky jerks and sucks an egg through a straw for so many others, for generations.' Brent sums up the sentiments of many when, in response to a question about whether he thinks seeds should be patentable, he replies:

> Does it benefit society? Only by enhancing the wealth of corporations. It is detrimental to the rest of society because it interferes with the free flow of unaltered seed. It also is detrimental to indigenous farming in which people have saved seed for generations and, all of a sudden, are not allowed to use their own seed. Seeds are life. No life form should be patentable. It violates the fundamental respect that all life deserves.

Patents are problematic not only because they limit the possibilities of saving seed, but because a patent is unethical in how it treats seeds – 'It violates the fundamental respect that all life deserves.'

Concerns about patenting were not limited to the restrictions placed on saving (and exchanging) seed or even on seeds as living beings, but extended to pragmatic breeding issues as well. As Denis explains:

> One of the problems with royalties is that it makes it less likely that the product, the actual introduced variety, will be used as a parent in some other new variety. So, it kind of restricts the family chain that can continue. And it means that some of that effort that went into breeding that variety, which may be a very good variety, it is being led to a dead end because there are property rights that restrict breeders from using that as a parent to make the next generation. That's a roadblock to continuing and keeping ahead of the changes that we need to keep ahead of.

Having freedom to operate in breeding is of serious concern, particularly when combined with already limited public breeding programs, and for Denis, IPR pose challenges even to more official efforts at producing public seed varieties.

Opposition to patenting seeds (or their parts) was almost unanimous, but a few savers offered qualifications that a reasonable return on investment for improved varieties might justify some form of intellectual property rights. Matthew, for instance, explains, '[w]hen someone works hard to produce a seed, they have every reason to expect some level of compensation for their efforts, whether that comes in the form of barter or monetary exchange.' Similar to others, he distinguishes between seed-related patents – that are 'too complex for the modern judiciary to handle', impede research, and foster a 'science for the betterment of the economy' rather than a public benefit approach – and plant breeders' rights that he feels could protect breeders' work in more reasonable ways. Savers like Matthew, supportive of some version of protection for seed varieties, applied caveats that current rights need amending to ensure that they not be applied for too long durations, too prohibitively or be too costly.

Genetic Engineering

In combination with IPR, genetic engineering offers corporations opportunities for expanding control. Further, genetic engineering, particularly when justified as improving food and food access, is to some seed savers simply a smokescreen for corporate pursuit of profits, facilitated by government policies. Phil explains:

> Way back, 15 years ago, there were all these people that were promoting bioengineered food, including our Canadian government, saying, 'Oh wow, this'll give us improved nutrition and better food all around, and save the planet', kind of stuff. And all these tests that they're revealing, cite absolutely nothing that their research was into bioengineered crops for better health and nutrition. It's all just bioengineered crops to withstand more and more applications of poison. And, they keep on, still saying that, 'this'll give us better food' and it's just bullshit. And everybody knows that.

For Phil, corporations and governments enrol genetic engineering in misleading ways, suggesting biotech provides 'better health and nutrition' when really they are meant to 'withstand more and more applications of poison.' A surveyed saver explains how a combination of GM crops, intellectual property rights and corporations propelled their resistance through seed saving:

> I first got into this because I was horrified to read about corporations putting patents on plants and then suing farmers if the wind blew their GM seeds onto the farmer's fields. The more I learned about what is happening to our food supply, the more concerned I became.

For both Phil and this surveyed saver, corporations, despite their claims, use GE to advance their agendas rather than improving food and agriculture.

Many savers indicate genetic engineering is problematic. A few savers object to GE technologies on the ethical grounds that such technologies interfere with seed living, but by and large articulations orient around the effects of GE and their techno-political relations – though these objections are not necessarily exclusive. Will joins up these two concerns in his explanation of his thinking around GE:

> It is negative because it separates nature into discrete pieces and is the antithesis of Nature's integration. It is negative because the people playing with this have no accountability and are motivated by Pride and Greed, characteristics not found in good farmers.

Genetic engineering, argues Will, is negative in multiple ways: GE cuts into natures, and the people using it are both unethically motivated and unaccountable. Dealing with GE organisms is the second most frequently listed issue in need of policy attention in the survey, with respondents including proposals for bans and for dealing appropriately with contamination. Every interviewee expressed concern with how GE issues have been dealt with thus far.

For many savers it is the technopolitics – the particular use and relations – of genetic engineering that are problematic, rather than the technique itself. For Phil and the surveyed saver quoted above, the close relations between GE and transnationals cooperating at agrifood reordering are objectionable. Other savers suggest that there may be benefits to the technology, but only if choice is assured and use careful – criteria not met thus far. Monica, for instance, submits:

> I can see that genetic engineering has some positive aspects but I feel that the consumer is entitled to be informed of the possibility so they can make their own choices on whether or not to make use of these, especially food, products. I personally feel that we all should have that right to choose.

Here, labelling would provide some choice, but Monica's preference is to grow her own food. Saving seed facilitates this choice. Denis explains his perspective through analogy:

> I always think it's really difficult to generalise when you talk about something like genetic engineering because it is such a big thing that is dealing with such fundamental parts of nature. It's sort of like discussing whether chemistry is a good thing or not. Certainly you can do terrible things with chemistry, but you can also do good things. And it's really a tool. It depends on how you use it.

For Denis, and several others, genetic engineering is a tool to be used (or abused) not an essentially good or bad thing. It is, in other words, a relational instrument. The problem of genetic engineering for these savers is that it eliminates choice and is used inappropriately.

For most savers interviewed and surveyed, GE seeds – with all they carry with them – are not worthwhile and are best avoided. As Brent suggests, the issue of genetic engineering is complex, but '[g]enetic engineering is a solution looking for a problem, especially as it has been used in agriculture.' Three quarters of surveyed savers strongly disagree that GE crops are needed for long-term food security. Phil's statement above articulates this sentiment, and Sandra echoes it:

> Start to tinker with Mother Nature and you will cause a chain reaction that usually continues into the unknown and eternity. Most of the time the damage is non-reversible and you have to continue with the band aid approach. Great stuff from the commercial aspect, since you are forced into buying only from them, but mostly they forget that technology and a single strain is in reality not a good match for all areas on this planet. Why use force when things are just fine within their own boundaries? Why create more of a mess in this world than it has already?

Here the concerns of growers articulated in Chapter 3 find resonance – GE crops spread, are not well understood, deepen dependence on transnationals, and do not meet the needs of growers. Further, seed saving and agro-ecological production offer better options for seeds, savers and their shared worlds. Ada sums the issue up in this way:

> All the propaganda that GM food companies put out about increased yields, better resistance to disease or pests, or putting an end to world hunger is completely misleading. If we were in better balance with Nature in the first place we wouldn't have the hunger, disease, and pest problems that we do today. We need to look at the whole picture of where we are, and why we are here, not just treat the symptoms.

In this context, many savers contend that GE crops are – counter to industry and government arguments – unnecessary. In this way, the biotech industry and facilitating governments misidentify the problem, and propose the wrong solution; instead, 'we need to look at the whole picture' to improve our shared worlds.

Contesting Commodification

If seed savers express anti-corporate sentiments, does this mean they reject seeds becoming commodities? Not necessarily. Instead, savers complicate and contest seed commodification. On the one hand, all savers resist and evade seed commodification by saving seeds, reusing them year-to-year, and by valuing seeds as more than mere consumables. In some ways this stance is a practical decision – it saves money, results in adapted seed, and so on. In addition, as seen in above statements, savers are also motivated by desires to disengage from corporate seed relations, including deepening seed commodification. Savers, however, are not exclusive in their practices.

So, on the other hand, many savers buy seeds as well as saving them, which might be interpreted as reinforcing commodification. As the survey indicates, few savers save all of their seeds. Most commonly for surveyed savers, saved seed accounted for 1–25 per cent of seed used in the previous season with 37 per cent of those surveyed indicating using 75–100 per cent saved seed. Interviewees did not specify sources and percentages, but conversations are suggestive of higher percentages in comparison to survey results.

While some savers rejuvenate supplies with new varieties obtained through exchanges, seed purchases also serve this purpose. Purchases also allow savers to try different varieties, and therefore different possibilities. And, through their purchases, savers relate with seeds as commodities. Even through their purchasing though, most savers complicate commodification – subsequently saving and sharing purchased seeds and evaluating sources and seeds with more than market logic. Accordingly, for most savers, seeds may be commodities, but they are always more than this. It is in this sense that savers contest commodification processes.

Savers purchasing seeds often make distinctions based on ethico-political practices of select companies within the commercial seed sector. Paul illustrates this by differentiating between marketing seeds generally versus the particular manifestation of seed corporations:

> I think marketing seed is generally a positive thing. It makes seed available, it ensures that viable seeds are preserved, and it makes a system of distribution possible. But large companies have a tendency to dominate a market; their brands and products get promotion, so their interest is served (even if the public interest is not). They sell whichever seeds they want based on their own financial advantage and because they can afford to advertise their own catalogue, they tend to limit what is available. So many of the older varieties of seed are eliminated from circulation; they may even die out completely and the gene pool be depleted.

Paul juxtaposes smaller, locally-based companies growing seeds for sale with the practices of transnationals and companies that redistribute corporate seed. The surveyed savers back up Paul's assertion, with almost all strongly agreeing or agreeing that they 'prefer to support small, local seed companies.'

The scale of operations, important as it is, is not the only consideration, however. While being local and small increases the chances that a seed company might be more sympathetic with the ethico-political stances valued by seed savers, in the end it is ethics rather than scale or location that matters. Hannah, for instance, clarifies,

> I think that marketing seeds by small companies would be better than marketing them by large corporations because if the companies are small enough, you can

know the grower and her ethics and she very likely feels very connected to what she grows, whereas large companies are faceless and are only in it for the profits.

Hannah values close relations and trust among producers and consumers, suggesting similar goals to relocalisation efforts. Quite frequently gardeners (or small-scale growers) are unable to determine the primary source of their seeds when purchasing from larger retailers, choosing instead other avenues for buying seed. Will makes the importance of ethics in this regard clear:

> Seed selling and marketing is not a bad or good thing and the size of the company is not important. The intention and application of this intention is the only determiner useable to judge the goodness or badness. Does the company encourage small, local seed saving and selection and promote positive agricultural practices and monitor themselves and dedicate themselves to supporting good farming? Then it's okay.

Scale here is subordinate to company practices: does the company 'encourage small, local seed saving' and 'good farming' as well as providing quality seed? It is these values that are most significant, rather than a company's size or location. Interests, influence, offerings and practices differ among companies and these factors affect savers' evaluations.

Savers who buy seeds as well as saving their own indicate an appreciation and preference for companies – usually small seed growers, often within a saver's region – that are supportive of seed saving and other valued growing practices (organics, biodiverse collections, etc.). Ada explains:

> The people that sell heirloom, open pollinated, organic, seeds are doing the rest of us a service by preserving our biological future. I don't mind paying more for that kind of piece [sic] of mind. ... I think of the money spent as an investment, in the future of food, and in my own personal seed bank (you only have to buy seed once).

Ada considers her seed purchases carefully, thinking of buying as reinforcing particular practical ethics and dealing with wider issues such as biodiversity and food security. Similar political implications attend these purchases as those mentioned in relation to alternative food movements, but saving pushes these relations further – beyond consumer activism. Buying from companies believed to be doing the right thing is complementary, for Ada and others, with continuing to save these seeds, after all 'you only have to buy seed once'.

It may seem counter-intuitive for small seed companies to promote seed saving, since the practice restricts future sales, but many seed grower-sellers share information on how to save seeds and are happy to hear that people are saving and exchanging their seeds. As Lena explains: 'Well, generally when I've talked to people who sell seeds, they don't have a problem with knowing that I'll want to save the seeds in the future. It's as if they are performing a service – making the seeds available.' Dan Jason (2005), a seed company owner and seed grower/saver, explained his perspective in a public address:

> I don't think that's [seed saving] going to jeopardise our livelihood because part of what we have to do as a seed company is keep on researching old and new varieties that are open pollinated and it's our job to tell people what's good, what grows best where. And as long as we're doing that we'll have people coming to us, new customers and old customers.

In fact, Dan spent much of his talk advocating that those with companies like his improve their skills and collections, that seed savers find ways of communicating across the country, and that everyone encourage seed saving. All of the other small company owner-growers who participated in this research expressed similar passions for seed and sustainable seed networks – passions that fuel their business – and though they certainly face challenges in maintaining livelihood, encouraging seed saving and selling quality seed were not seen as incongruous.

In seed savers' understandings, commodification of seeds is not necessarily problematic, nor is it permanent. This valuation of seeds as more-than-commodity is different from that promoted by transnational corporations wherein seeds become commodity-technologies – disposable, ownable containers to be profited from year-after-year. Instead, the form and ethics enrolled within commodification and marketing are valued under particular circumstances. Market logic may be part of seed-saver relations, but it is only part. Seed saving here has less in common, I think, with neoliberalisation's commodification (discussed in Chapter 3) than that of Gibson-Graham's (2006, 2008) diverse economies, which include market, alternative market and non-market exchanges. Moreover, becoming a commodity reflects only one possibility in the life of any seed. Appadurai (1986) suggests that things – such as seeds – have social lives of their own, and as they move through this social life their matters and relations, and our valuations of them, alter. As we have seen, seeds may be commodities one moment and in the next become art, food, gifts, saved or even several of these at once. Seeds, therefore, are always more than just commodities, and in saving, commodification becomes contested.

Creating Self-Reliance

Self-reliance, as savers express it, is directed more to creating and reinforcing alternative seed relations than to direct resistance. While speaking of self-reliance, most people concentrate on ensuring diversity in collections and ecologies, maintaining household and community resilience and food security, and keeping seed in the public domain. This more proactive approach is then supplemented by comments that people save seed 'to be free of capitalistic corporate greed' or 'to take back our heritage from seed companies.' So while self-reliance is contrasted to dependence upon corporate seed ordering, the emphasis is on building seed saving practices and networks.

The desire for, and attainment of, some level of self-reliance is one of the primary motivations for saving seed. According to Kneen (1993: 119) self-reliance in food networks is not about individuals creating for themselves enough to eat; rather, self-reliance, 'means relying primarily on those people with whom one lives from day to day and on the resources at hand, rather than being dependent on outsiders and external resources. It means carrying on external economic relations on the basis of equity and mutuality, not exploitation.' For Kneen, therefore, self-reliance is not an individual achievement, it involves collectives. Further, he indicates that local relations provide a base, but that connections with others further afield are sought when necessary and oriented by ethico-political principles. Similar notions prioritising local relations and social justice principles have also evolved as part of food sovereignty movements.

Using this broader conception of self-reliance, seed savers speak of strengthening individual, household, community and national achievements employing terms like autonomy, control, freedom and self-sufficiency. Surveyed savers indicate strongly that seed saving improves their self-reliance and positively impacts their communities; 83 per cent agree or strongly agree that 'seed saving has improved my self-reliance' and 88 per cent agree or strongly agree with the statement that 'seed saving positively impacts my community.' Seed saving is not trivial; rather, it relates directly to sustaining lives and is of common concern for people around the world. As Irene explains:

> It's all about control. Who has control over – you know it sounds silly, but who controls the seed controls the people. In lots of ways and lots of places in the world people's ability to feed themselves year after year is still very fundamental to the survival of the people. It's a survival issue.

This survival involves having access to, and abilities to grow, seeds that are appropriate for local ecologies and cultures. Seed saving, therefore, allows growers to evade deepening relations of dependence and to have some control over shared futures. Global food sovereignty movements make this clear in their calls for local control of resources, as do the seed savers with whom I spoke.

Seed saving is articulated as providing a possibility for freedom and security through self-reliance that is missing from relations with, and in, corporate seed ordering. Carl puts it this way: 'With the current day amalgamation of seed companies worldwide, and the disappearance of our seed diversity, seed saving is not only important to me, but crucial to my sense of security in an ever [more] industrialized and globalized world.' Here the problems to be dealt with – corporate concentration and loss of diversity – are addressed through saving seed, providing Carl with a 'sense of security'. Adam, a saver with a small seed company, offers a slightly different perspective by arguing that 'local, independent agriculture must have seed saving as an integral part of it, or there will always be a dependence on outside/foreign sources of seed, that may not meet the needs or desires of local farmers.' Seed saving, for Adam, is foundational to sustainable and 'independent' agriculture. Both of these savers illustrate how political and ecological concerns inform their practices, and how seed saving provides, for them, a means of addressing these problems by remaking seed relations.

The importance of self-reliance relates not only to food and farm security, but to control of production processes. Dara, for instance, expresses her concern for food and its connections to saving seed by revealing: 'It is really important to me to save and provide as much food as possible for me and my family to eat and to be growing it naturally. It's not only to be self-sustaining, but to be able to control how it's grown. That's really important to me.' Susan also finds reassurance in self-reliance through seed saving. She says:

> partly I resent having to fork out money to pay for seeds to seed companies when I could do it myself. And I also like to know that, that I'm not dependent upon anybody for what I need for my food. ... like if I knew that I had to go out and buy all my potatoes every year to plant it would really bother me. But I know I've got a basement full of seed potatoes that I can, I'm all right, you know?

For Susan, thrift and independence link together in her saving through her concern for food and seed. Having alternatives – the seed potatoes in the basement – serves multiple sensibilities. The quest for self-reliance is, in part, a quest for control to ensure that the needs of savers (and some related others) are met in better ways. In explaining the origins of his seed saving, Brent says:

> My seed saving has evolved partly from my desire to be self-reliant, partly from my innate sense of thrift, and partly out of my education about the agricultural system. I enjoy being capable of meeting as many of my own needs as possible. This gives me a stronger sense of security and stability.

As Brent states, echoing many participants, there is satisfaction in recreating self-reliance – in addition to other benefits. Susan sums it up very simply: 'I like the feeling of being able to rely on myself and friends and neighbours, and not have to buy seeds from seed companies.' The self-reliance fostered through seed saving, as these savers demonstrate, not only resists corporate control, but builds feelings of security and satisfaction for those involved.

Fostering Collective Belonging

Much seed saving is done by individuals who sometimes feel isolated in their pursuit, but seed saving, like other practices working at re/creating alternatives, does not happen alone. The connections between people and seeds are evident from the time one begins saving seeds; however, I was reminded of the sense of belonging among seed savers when I walked into the Heirloom Seed Sanctuary in Kingston, Ontario. I had been living in Kingston for a few months and in my internet trawling came across an announcement of a series of seed saving workshops happening just up the road. I had been saving seeds for a few years, but in Kingston I had a bit more space and was planning on expanding my efforts in the next season. As a novice, I wanted to increase my understanding of how to do saving and growing well, and maybe try a few more varieties. As a researcher, I was interested in how the workshops would work, the place where they were happening – a 'sanctuary' for seeds – and the ways in which other savers work and think with seeds.

Fig. 8.1   Considering seeds at the Heirloom Seed Sanctuary

The first workshop I attended was a useful for me because of the information shared, but mostly because of the connections I made with other savers. While I lived in Kingston I returned for several workshops, a couple of drop-in visits, and volunteering on a few workdays and public events – including the Sanctuary's well-known tomato tasting day that brings in people from across southern Ontario. I bought and exchanged seed and shared ideas with people there, and our relations led to the organising of Kingston's first Seedy Saturday (which continues). When I recently returned to the Sanctuary for a visit I was happy to see a thriving collection of seeds and seed savers. Community-based seed libraries and sanctuaries, like this one, exemplify savers' efforts to build upon their ongoing informal relations by expanding learning, collections and connections.

For seed savers the 'self' in self-reliance, as suggested earlier, is not an isolated individual but one immersed in collective processes of provisioning and practice. Most often, saver collaborations are informal – conversations among friends, family, neighbours – but in other cases they take more organised forms like seed fairs and exchanges, workshops, internet networks, memberships in seed sanctuaries or other organisations. All of the seed savers I spoke with engage in various collective activities related to seeds and find these sharing activities integral to their seed saving (also explored in Chapter 6). Paul illustrates the importance of cooperation through an example of how saving happens within his community garden:

> [T]he nature of seed saving makes it a cooperative activity. I have a plant list of the allotment garden. There are about 30 varieties on it (vegetables & herbs, not flowers). If I had to rely only on my own seed collecting, even that small area would become a major chore. Sooner or later I would fail to collect a variety and lose it. But if five people collect five different varieties, or ten collect merely three, the task soon becomes manageable. It makes seed exchanges look very desirable.

In order to save several varieties and reduce the risk of loss, Paul relies on other members of his garden to contribute to shared seed saving. And he extends his example to seed exchanging in general, suggesting that seed saving is collective practice.

Beth explains one of the ways in which she learned about seed saving as part of her social relations with new neighbours, recalling:

> I happened to move into a house where my two closest neighbours are very serious gardeners and one of them does a lot of seed collecting and plants a lot from seed – like she'll start a clematis from seed and whatnot. So, it's just 'Yahoo!', a wealth of information and I think that inspired me a bit too. Someone close by that's doing it too and knowing that I can ask a question, you know?

Cooperation among seed savers occurs through the exchange of seeds, but also through a sharing of knowledge and inspiration. Beth's neighbours have contributed to her saving in both ways, fostering the development of her collection and knowledge. Further, Beth has followed their example and continues to share her seeds and knowledge with people she encounters that show interest in gardening. Beth continued speaking about the sharing she now does with others, reflecting that 'every time I give away a plant it's just giving away a little part of what you've created on to someone else to grow in another place. I guess that's what it is really – just a bigger sense of community.' Individual savers do not disappear in the examples offered by me, Paul and Beth, but the isolated individual of neoliberalisation is countered as savers function necessarily as part of collectives.

For some, sharing relations were predominantly, though not exclusively, local. For example, Pat describes her connections this way:

> I do feel that culture and community play an important part in seed – and plant! – saving. They connect us almost in a kinship way – as well as friendship and more practical reasons – as we share them and then see them growing elsewhere in the community. Also, seeds/plants grown and shared in community develop into community cultivars particularly suited to the local microclimate and growing conditions.

Here, community operates within a relatively nearby geography, with seeds serving as a connection between savers as well as becoming part of a socio-ecology as 'community cultivars.' While the local does receive much attention by savers, the idea of collective belonging is not exclusively applied at that scale.

Some savers emphasise a sense of belonging based on shared knowledge and practice with other savers regardless of location. For instance, Alice reveals: 'I give away seeds to my friends', laughingly adding, 'and if we have only just met and you mention you like to garden, you qualify as my friend.' This feeling of connection is expressed by many savers in different ways – other savers become kin, friends, part of a community, if only for a short time. Articulating more dispersed relations, Irene describes several ways in which she engages with others:

> I belong to Seed Savers USA and Seeds of Diversity Canada. ... [I connect with other savers by] simply reading the magazines that they both produce and by riffling through yearly seed exchange catalogues that come. Obviously, that's not connecting one on one, but I'm hearing about their stories and reading about their passion. I love those magazines for that reason. And I also attend my local Seedy Saturday, where I met you, and I love that event. And I just have a few friends who are seed savers.

Irene combines face-to-face informal relations with friends and those she encounters at Seedy Saturdays with more at-a-distance connecting to savers through reading, seed exchanges and organisational memberships. Her experiences with others – even when mediated through magazines, organisations and physical distance – are not unattached; rather, she participates affecting others and being affected in turn.

Lena elaborates on her own sense of a seed saving community saying that when she goes to Seedy Saturdays 'it seems like we're all speaking the same language, so I definitely feel a connection that way.' She continues:

> To know that other people are doing the same things and just – it's – it's a community that understands the concept and also a community that I can share seeds with. When I send seeds, I usually write a little note about it or about how to grow it, how to save the seeds, and I really appreciate getting that information when I order seeds from someone too. Even though I'll probably never meet that person who's across the country, for that moment, for that exchange, it does feel like we are friends in the seed saving sense.

Despite the geographical distance between her and those with whom she shares – sometimes savers living on an opposite coast over 6,000 kilometres away – Lena feels connected with other savers through her sharing of seeds and practices. To share with others and to be shared with in turn – however fleeting a particular exchange – is part of the pleasure Lena, Alice, Irene and others gain through saving seed. These connections build and rebuild a sense of engagement and belonging with other savers.

Journeys of past savers and seeds are also appreciated as part of these collectives – an integral part of how seed saving is made possible, and one reason it continues. The problematic seed-saver histories wrapped up in colonialism and biopiracy debates, though mentioned, received limited attention in the interview conversations. Instead, focus fell on present saving efforts aimed at continuing traditions within reconstituted seed networks. It is unclear whether in such conversations savers were avoiding difficult questions or were simply centring on ways and places they felt they could engage. In either case, current seed saving is understood as part of previous efforts as well as part of recreating possibilities for those to come after.

The sentiment that seed saving bridges past, present, and future finds expression in Jane's comment that '[s]eed-saving connects me with previous generations because the seed I am planting is only available to me because people in previous generations saved it and replanted it. By doing the same, I am continuing the cycle – so that future generations still have access to them.' Hannah articulates this feeling as well when she states that, 'Seeds link people of the past and people of the future in that they can create realities of the past for people in the present and the future, if people choose to keep those realities alive by keeping the seeds viable.' Nazarea (2005) develops such connections of memory and seed saving, arguing that the remembrances and continuities accompanying saving challenge the 'culture of forgetting' upon which corporate seed ordering relies. As Jane and Hannah indicate, the relations of present seed saving spread beyond present time into the past and future through relations that manifest in common practices and seed embodiments.

The sharing of experiences, knowledges and seeds serves as a way for seed savers to contribute to their various networks of belonging, but also to seek support and inspiration to continue saving seed. Rachel offers that sharing of this kind 'serves as a connective element between people. It's a gift that grows!' This connective, collective character of seed saving is sought out and built upon by savers as they go about saving seed. This politics of belonging includes reconstituting connections to past and present relations as well as expanding future possibilities.

Cultivating Alternatives

Savers value the differences their saving makes in wider worlds, and continue saving – at least in part – because saving supports and makes possible alternative ways of living. Perhaps one of the reasons that relocalisation efforts are so popular is that when considered at global levels food issues can seem overwhelming and impossible to alter, while actions and their effects feel possible and are experienced in everyday living. As Allen (2004: 65) argues:

> The work of developing alternative practices and institutions is a real and immediate point of engagement for people as they go about their daily lives. ... Food-system alternatives can create and connect economic and social spaces and establish new models that engage public concerns about community, social justice, and environmental sustainability.

Most seed savers with whom I spoke would agree. Phil explains: 'This [seed saving and growing food] is real, this is what we do every day. Food is really part of us. And, it's not a vicarious experience. It's something that is part of us.' Through their saving people are able to engage with things, and the differences made by their practices are lived.

Savers express their commitments in relation to different issues including food, the inherent value of seeds, the resilience of adapted seed, and so on. Jane seems slightly embarrassed as she explains why seed saving matters to her, but her conviction is firm when she says: 'I felt by saving seed, that I would be helping to save the earth! I know, it sounds like a cliché – but that's truly how I felt, and how I still feel.' Jane is particularly interested in conserving biodiversity and often chooses rarer seeds with interesting histories, saving and exchanging them and their stories. She goes on to explain that saved seeds have better possibilities of continuing, that with their continuance humanity's chances improve, and that by using adapted seeds and organic processes her saving matters in wider ecological ways. Carolyn articulates her own interests by revealing: 'Seed saving has empowered me' because 'I have a choice over the quality and types of food I eat.' She adds, 'It has also made me much more aware of the amount of labour and expertise that growing one's own food requires. ... now I have a greater appreciation of the value of food.' As all of these seed savers suggest, seed saving offers people a sense of living in better ways – individually and as part of collectives – and supports a sense of efficacy.

The importance of acting in the world is emphasised by many, and Brent puts it plainly when, showing me his seed collection after touring me around farm he worked on, he says that seed saving is '... a political act. And to me it's more important to do than to preach or instruct. Those are important things, but it is more important that I just garden and save seed than it is to write a book about gardening and saving seed.' I am sure I blushed at the time – maybe he did too, after all, I was there doing research to write about seed saving. But his words stick with me, reminding me that research and the passionate witnessing I attempt is valuable, but that seed saving makes differences not only by evoking thought – by presenting opportunities for rethinking relations and worlds – but by enacting alternatives.

Seed saving offers possibilities of different ways of living together, and savers develop a sense of community contribution through their engagements. Sharing knowledge and skills provides this kind of sense of contributing to wider communities, as Carolyn reflects: 'It makes me feel good that I possess skills and knowledge that few people still do and that I can share it with others. I am carrying on a tradition which has been largely lost to Western society, especially amongst city dwellers.' By sharing what appears to be rare knowledge with others Carolyn feels she is continuing traditions, forming new relations with people, and helping to re/build desirable food and seed networks. For others, donating seeds themselves provided this sense of community commitment. Natalie and her husband, for instance, feel that they have 'helped the community by donating seeds to families, family organisations and seed exchanges because gardening decreases the debt to income ratio – improving the local economy and its nutritional health as well.' This sense of participation in, and contribution to, communities is important to seed savers. Further, such involvements bring abstract desires to improve shared worlds into ethico-political engagement – as conserving agro-biodiversity and restoring ecologies, running workshops to share knowledge, sharing with others produced food and seeds, among other things.

Reconstituting agrifood orders is one way in which savers suggest they contribute to better lives for individuals and communities. Anna and Phil both describe a particular interest in soybeans (which you may recall from Chapter 3 account for much GE acreage worldwide):

> I would like to grow and save soybeans again – I think they are one of the most important things we can be organically growing and saving. There are so many varieties, many from Japan, that are very delicious compared to the soybeans we are used to seeing on our health food market shelves. It is increasingly harder to find organic seed for soybeans in large quantities (for an acreage planting, for example). (Anna)

> [T]here are actually a lot of other really great soybeans that aren't like the yucky ones that you find in the stores, the yellow ones that are pretty much like the same variety that Monsanto has for their roundup ready soybeans. There are brown soybeans and black soybeans, and soybeans that are so – as moist as much as any baking bean, even more so. They're just so light and delicious. They're just not like what anybody has ever experienced. Yet nobody grows soybeans just for eating as a dry bean. So, on and on, there's just story after story like that. All these beautiful crops like quinoa and amaranth that people have been enjoying wherever they're native, now they're told that they're an inferior food by transnational corporations and don't sell the seeds to grow them. But they're amazing foods. ... And yet, they're not in our market 'cause, as with a lot of crops, they don't stand up to being grown on millions of acres at a time with harvesters that are, you know, 400 feet long or whatever. (Phil)

For both Anna and Phil, saving soybeans makes differences to diets, tastes and health in ways that run counter to dominating seed and food orders. Further, Anna maintains that finding organic seed is increasingly difficult, while Phil points out that some soybeans (among other food crops) don't suit industrial agrifood production practices. Anna and Phil each convey changes in material and practical relations that come with soybean saving, and the importance of ensuring the continuance of these differences.

Savers sense the differences seed saving makes, which fosters feelings of accomplishment and empowerment. Lena simply states that seed saving 'makes me feel that I'm doing something good for the world.' And Natalie discloses that seed saving:

> makes me feel important. Like my life makes a difference. That I contribute. That I can influence others through my actions, and remind them that grandma's seed could be more special then they thought. That I have something to give to others.

Seed saving matters not only to seed savers' personal everyday lives, then, but is meaningful in relating with others and reforming worlds. As another example, Gail explains her understanding of her own seed saving and its enrolment of her commitment to those she understands as related others:

I am sympathetic to the cause of indigenous farmers all over the world to be able to feed themselves by saving their own seed. This self-sufficiency is being seriously eroded by globalization in agriculture. So I have begun to be involved actively in these issues. In short, I think as a young person, I was influenced to think, 'Think globally, act locally.' And seed saving to me, is an embodiment of that principle.

At another time, Gail reiterates, '[w]hat was once, my own private project of saving seeds ... has now become a focus for much larger international issues.' As Gail illustrates, whether people become savers because of ethico-political motivations or not, through seed saving many savers come to consider other actors in the food system more deeply and fully, which, in turn, reinforces their commitment to saving seeds as a means of expanding future possibilities.

Through their everyday practices with seeds then, seed savers engage in reconfiguring power relations, moving beyond their fields and gardens. Ada connects seed saving practices explicitly with power and control, saying: 'Seed saving puts the power back in the people's hands, and it is so easy, so why not do it?!? Seed saving just makes sense!!' Saving seed for Ada is clearly part of efforts to redress neoliberal agrifood ordering, facilitating savers' capacities to reconstitute alternative possibilities through a seemingly small act of everyday living. In this way, reflecting upon abstract and complex agrifood ordering, for savers the question becomes not a shrugged 'awh, what can you do?' but an interested 'what can I *do*?' right here, right now to make a difference. Many savers value greatly this aspect of seed saving, suggesting that saving seeds is understood as one way of 'doing their part' to make individual and collective worlds better to live in.

The practice of seed saving is always ethico-political in its enactment and relations, but some savers articulate their practices specifically as means to engage in changing seed–people ordering. In considering what is being accomplished as seed saving is enacted, combining reading for creativity and for opposition, this chapter has illustrated how saving and savers participate in resisting corporate control, contesting commodification, re/creating self-reliance, fostering belonging, and cultivating alternatives. In these considerations we have seen that seed saving occurs and savers respond in multiple ways – for instance, rejecting corporate seed ordering but not necessarily commodification, challenging some forms of IPR more than others, or making demands of, being frustrated with, and ignoring governmental policies. In this way, seed saving relates with corporate seed orders, and its supporting relations, in complex ways involving contestation, evasion, ambivalence, and more. Seed saving endures, and its continuing reconstitution by those seeds and savers practicing together provides opportunities to review and remake our shared worlds.

Chapter 9
Concluding Thoughts

Fig. 9.1   A selection of dried beans
Source: Cypher 2012

Returning to Beans

I started this book with a story of experimenting with beans as a child, and here I return to beans but as part of more current relations. Aside from recognising differences in form – shape, colour, texture – what else might you say about these beans? There are four varieties – some you might recognise, others perhaps not. Are they open-pollinated or hybrids? It is unlikely that they are genetically engineered – beans are not high on that priority list. Are they edible? Are they organic? In what conditions will they grow well? How did they get to here – via transnational subsidiary or contractor, local seed grower, passed down through generations? How do they taste? Have they been gifts, commodities, accessions, property, immigrants? Maybe you can identify particular varieties that would give clues to their paths and dispositions but nothing for certain.

The pictured beans are not from my collection. In fact, I distributed my collection among friends, family, freezer and compost bin before moving to Australia – where they would not have passed biosecurity inspections. I bought these seeds for a friend while on a trip to Canada and attending a Seedy Saturday. They come from a small company run by a woman whose seeds I came to know through the research for this book. Though I have not saved these particular seeds, I have grown and saved the varieties in the past. I was never a big dried bean eater, until becoming a seed saver – some of these varieties (among others) convinced me. I thought these beans would grow well in my friend's area, and she would enjoy their different tastes, forms and growing habits.

Three of the varieties pictured are not particularly rare, but neither are they the most common. The Blue Jay beans (dark bluish with beige marks) are relatively rare, and recently reintroduced by Seeds of Diversity Canada. All are open-pollinated bush beans (my friend did not want pole beans given her windy space and preferred timing for produce). None are hard to grow, harvest or save – suitable for her (and her family's) purposes. All are great in soups and baked, lending their own flavours and textures. I am told the Orca (black and white) and Blue Jay beans are also good fresh. Each is a heritage variety; the Jacob's Cattle beans (mottled white and maroon) are believed to date back to the 1700s in North America, and the Kenearly Yellow Eyes (white, with yellow-brown 'iris' around a white eye) are traditional for some East Coast baked bean dishes.

As part of my friend's collection, the descendants of these beans will go on to be eaten or saved, adapted and regrown by my friend. Or they will die from disease or neglect. Perhaps they will be passed on to another saver. These beans are lively, changing and affective, working with savers interested in continuing to save (and eat!) them. My show-and-tell conveys some of the stories of these beans and their people, but I might have acquired beans in other ways.

Instead of these beans, I might have bought beans to eat from the supermarket, or to plant from a garden centre. This is a privileged position to be in, and taking this option would have been less involved than saving seed. Going to the supermarket or garden centre would also have offered different options. The pictured varieties would likely be absent – only on a couple of occasions have I seen Orca and Jacob's Cattle beans at specialty garden centres, and never have I seen any of these varieties in supermarkets. Broad appeal, long food storing, uniformity, easy mechanical harvesting and expected chemical inputs or defences commonly come with commercial agrifood; other varieties are better fits for that ordering. If I was looking to grow beans, I might not have bought seeds at all – choosing instead to purchase plants ready to go into my garden. Perhaps I would have missed making the link that beans – like many seeds – are simultaneously food and seed. Certainly the opportunity to learn with and respond to seeds through saving would have been lost.

Relations with/in Seed Saving

We come to know worlds through our enactments of them – training our senses, emotions and thinking to recognise and respond to differences and similarities, limits and possibilities in those worlds and ourselves. Throughout this book varied ways in which people and seeds shape and are shaped through practice have been presented. Each set of seed-people relations involves ways of thinking, feeling and doing; in seed saving one learns and responds to seeds differently than in genebanking accessions or facilitating international seed trade. We have seen how such learning through saving seed allows savers chances to gain material and emotional benefits, contribute to collective change, and experience shifting influence that helps question valuations of seeds as mere natural resources. As part of saving seed collaborations, seeds become complicated things with lives, dispositions and contributions of their own. In this view, seeds and savers make differences, to each other and together.

Close attention to seed saving has highlighted experiences of a particular human-nonhuman relation and the ways in which that practice becomes meaningful for those involved. Though much of the practice and its meanings may be shared, seed saving cannot be understood as simply serving one purpose, offering one experience, or meaning the same thing for all savers in each moment. There are times when one simply wants to eat some beans, already grown and saved. At other times, concerns are evoked about seeds' journeys on the way to a field or garden, the pests and beneficials, corporations and governments, food security and justice, and more. As savers continue to save seed, they wonder and despair, feel powerful and insignificant, offer support and resistance, among other things. The affects and practice of seed saving are not unitary or staid, but they are recognisable.

Seed saving has been demonstrated to offers ways of living with seeds different from those achieved through corporate seed ordering. Moves toward lab-based breeding, expert knowledge, intellectual property rights, genebanking and genetic engineering have been entangled with the building, maintenance and expansion of a corporate seed order through neoliberalisation. The multiple practices that come together in this kind of practicing are both similar and different to seed saving; similar because, in laboratory or in field, seeds are grown, monitored, distributed and stored, and experiments are conducted. In some cases, seeds may also be multiplied or regenerated. But the specifics – the who/what involved, and how and why it is done – vary. Each practice is situated, and matters differently – in particular enactments and wider worlds.

Different seed-people relations bring with them particular possibilities and limits. Illustrations have included how saving brings death into growing spaces while immersing savers in seed life cycles; genetic engineering allows recombination of genes from reproductively incompatible organisms; genebanking maintains select seeds, if imperfectly, for long periods; intellectual property rights facilitate privatisation, and so on. In addition, these practices enact inclusions and exclusions. Within seed saving, for instance, some savers keep many seed varieties while others save just a few, and in any enactment of seed saving some seeds are lost – to eating, disease or other causes. In other seed-related practices, Terminator technology works especially hard at excluding some seeds from being saved, while patents reinforce only some ways of knowing seeds and authorise select activities. Different ways of growing, breeding, banking or regulating alter the options for seed-people relations and the ethico-politics that come with them. Farmers' rights, participatory plant breeding, molecular marking without modification of organisms, development of small-scale seed growing businesses encouraging saving are just some examples of ways of doing things differently that have been noted

In this book's chapters, varied seed-people arrangements have been explored, each accompanied by their interactions with seed saving and its practitioners. Some encounters bring clashes, abrasions or contestations – savers resisting corporate control of seeds for instance, or farmers' rights existing in friction with intellectual property rights. In other cases, participants remain indifferent, carrying on as they would otherwise – seeds reproduce and savers persist despite legal constraints, seeds die despite genebanking efforts, or policy changes proceed without public engagements. In still other meetings, practices may work together, supporting and reinforcing each other – such as genetic engineering's allowance of deepening commodification and corporatisation, or genebankers working with savers on regeneration or breeding and exchanging seeds in the process.

These relations are not simple. Corporations, for instance, display advantageous ambivalence as they move between regulatory realms; on the one hand, promoting GE seed as contiguous with 'conventional foods' when seeking safety approvals, while, on the other, claiming seed (or seed parts) as invented products for intellectual property allocations. Savers, too, illustrate complex participation – contributing to, rethinking and counteracting seed commodification for instance. Each set of seed-people relations and each practitioner encounters other enactments and arrangements, establishing dis/connections.

A fuller understanding of seed politics comes from exploring experiences of people's living with seeds and saving, and how these experiences are part of learning and responding to seeds and other seed orderings. Figure 9.1 and the story of those beans reminds us that how we interact with nonhumans involves not only those relations, important in and of themselves, but possibilities and worlds beyond them – connecting everyday practices like saving seed with food and supply chains, culinary and production traditions, and ethico-political valuations. Seed saving both remakes and is remade through practice and in encounters with other seed-people practices.

Practicing Change and Changing Practices

Saving seed has not only changed the way I grow things, it has changed my life. (Anna)

They're not going to stop all the 8 million Canadian gardeners from saving their own tomato seeds. Probably. Because people will be so up in arms, maybe that's the most revolutionary thing to do – to get people saving their own seeds. (Phil)

For these seed savers, among all the others with whom I spoke, seed saving forms a vital – in the dual sense of lively and valued – part of who they are and how they engage in the world, and each argues (in different senses) that seed saving is transformative (of individuals and of societies). While perhaps not a revolution, saving seeds contributes.

Seed saving provides opportunities to engage with seeds, rethink possibilities and reconstitute worlds, but whether and how to take up these possibilities remains in question. Making these decisions comes down to responding to two basic questions: what should be accomplished through seed-people relations? and, what are we willing to do in support of such relations? How, in other words, do we – as individuals and collectives – move toward more responsible and sustainable futures? I have argued that seed saving can play a valuable role in reconstituting our worlds in more positive ways, and that savers already go some way in cultivating these futures. More might be achieved to connect wider publics with seed saving, but the practice is gaining attention in the popular press (particularly as part of food movements). Such attention provides a chance to not only advance the practice 'on the ground', but to connect such moves with policies and technopolitical orders. We can push and pull our worlds and selves into new forms through practice, though this is not always pleasant or easy – as the frictions among seed-related worlds illustrate.

Three ways ongoing efforts to facilitate seed saving might be enhanced have been raised throughout the book, each of which might be applied more broadly to altering practices and their orderings. These ways include changing the participants involved, the relations within a practice, and/or the encounters among worlds.

Changing Participants

The first way of altering arrangements involves changing who/what participates – in seed saving or some other practice. In basic terms, where there is no open-pollinated seed, and no willing savers, seed will not be saved. In this way, altering the 'things' available, changes possibilities. Corporate decision-making about which seeds to offer on the commercial market illustrates this point clearly, as do the efforts of savers to continue growing and sharing diverse varieties like the beans at this chapter's beginning. Just as low-level presence policies would expand markets for GE seeds, so a new greenhouse might extend the season and increase the likelihood of seeds maturing.

Changing practitioners in ways that boost seed saving evidences in initiatives to develop exchange networks, offer seed for sale through small-scale companies, recruit new savers while retaining existing savers, and even becoming a saver oneself. The seed available also needs to adapt to the growing conditions and cultures in which it lives. Policies supporting such measures might include exempting smaller amounts of seed from registrations, providing funding – as was done for the biotech sector – for initiating and developing community and national seed networks, supporting public breeding of open-pollinated varieties to be released without limitations on saving, or publicly promoting seed saving as contributing to food security, conservation or other related issues.

It is not enough, however, to have seeds and people available – the practice itself must be maintained and supported. Support may be garnered from various things: sustained electrical service might allow cool rooms to keep seeds longer, a book encourage a particular method, or a variety enable a conversation about saving. Further, these facilitating things must not only exist but be known and used – a book's advice unread, for instance, remains excluded and untrialled. In order for saving to thrive, networks for sharing seed and knowledge need to be maintained, improved and expanded. To achieve this, practitioners must be motivated to engage, and continue that engagement. Understanding how and why a particular practice becomes meaningful provides insight into how it might be fostered. In relation to seed saving, we have seen how it (among other things) evokes satisfaction, offers resistance, provides material benefits, fosters joy and efficacy, provokes curiosity, demands skillful and care-full attention. The future of seed saving, at least in part, depends upon deepening and spreading the practice's materials and attachments.

Changing ways of thinking to recognise practitioners in more complicated ways may also be required. We can think of how seeds exist in saving as commodities, biodiversity, food and living creatures, or how savers act as consumers, activists, rights holders, producers and conservationists. In this sense, practitioners must be understood as complex beings. Including nonhumans as practitioners, or collaborators, with their own lives and agency offers and demands different relations than does using them as natural resources, owning them as property, or profiting from them as commodities. Working to understand and reform practitioners' various enactments and motivations is an ongoing challenge, one that requires acceptance of multiplicity accompanied by ethico-political engagement.

Changing Relations in Practice

A second means of changing our shared futures occurs through shifting relations among those involved. In other words, the experience of the practice and how it comes together may change. We have seen, for instance, that certifying seed has become a means of expanding commercial markets through altered connections with market valuations, industry standards and authorisers, and government priorities, but it was not always thus. In relation to saving practices, we can recall how saving seed with conservation in mind requires different methods in comparison to adapting or experimenting with seeds, while community seed banks (along with seed libraries and sanctuaries) provide innovative ways of obtaining seed and connecting with other savers.

No one actor controls how a practice comes together, but some prove more influential than others in particular arenas and at specific times. The strategy of forum shopping employed by some countries seeking strong intellectual property rights provides an example. The concept of degrees of freedom helps reveal ways in which relations of power inhere in practices. Recollect, for instance, the interplay among seeds and savers during different phases of saving – when savers are forced to wait for outcomes, when seeds are selected out, and so forth. Initiatives to train savers and adapt seed are suggestive of ways in which changing practical relations might serve to improve seed saving. Though constrained, relations among practitioners are dynamic.

Categorisations of seeds, savers and saving can reconstitute practices as well, and often such categorisations underpin relations of power, facilitating some relations while inhibiting others. This book has offered several exemplars, showing how divisions among nature and culture, objects and subjects, and expert and lay knowings serve particular interests and activities. For instance, facing seeds as plant genetic resources for food and agriculture and furthering this approach by turning seeds into accessions in genebanks not only refigures seeds and savers in ways that ignore their complex relations and histories, but also facilitates commercial breeding cultures, increasingly controlled by corporations. This approach includes defining seed as part of nature (as opposed to culture), and as object (versus subject), useful (compared with inherently valued), and knowable through science (rather than learned through saving practice). Delineations in intellectual property rights, particularly through patenting, display a similar approach. As a more straightforward instance, the exclusive attribution of purity by industry to certified seed as opposed to saved seed, combined with campaigns to educate growers and contractors, works toward seed revaluations with impacts on the bottom lines of both savers and the seed sector. Corporations have proven adept at responding to and influencing such divisions; seeds may be maneuvered across nature-culture boundaries established in different orders – becoming inventions, natural resources, or equivalent to conventional foods but they remain commodities in the service of corporate profits.

Unsettling claims and categorisations, and the activities they facilitate, is part of how different ways of relating might be encouraged. Ag-biotech represents their approach – seeds as commodified, growers as consumers, saving as anachronistic – as both desirable and inevitable. But such refigurations are disrupted in seed saving, becoming only one possible relation within complex entanglements among savers and seeds. Saving, in this view, becomes not anachronism but vital practice. Seed–people orderings are not only dynamic then, they are contestable.

Changing Encounters of Worlds

The third means of changing seed–people relations works through altering the ways in which different practices and their orders encounter one another. Relations among seed-people worlds are diverse and multi-dimensional, moving beyond common dichotomies of collusion or opposition. As demonstrated, practices (and their resulting worlds) may affect others by reinforcing, overtaking, resisting, ignoring, fostering, constraining, facilitating, and more.

In some cases, orders cannot accommodate one another (or do so at their own peril) – as seen with disputes between GE and organic growing, patents and seed access, Terminator use and seed saving. In other moments relations are less clear-cut, pointing to the multidimensionality of worlds-in-relation. In policy worlds, governments may rescind seed registrations over concern for lost markets, be unwilling to include economic risk criteria as part of approvals, and advocate low-level presence – each time connecting differently to farming, ag-biotech, and international trading arrangements. In similar fashion, various practices come together at different times with saving. Conserving diversity in fields and with genebanks, food provisioning for individuals and wider communities, ethical concerns about seed flourishing, and political protest of corporate control and patents are just a few examples of practices that entwine with saving, not always in complementary ways but always relating with the practice and its meaningfulness. Seed-people worlds relate in multiple and complex ways.

As the strength, or durability, of each enacted order varies so does its influence. Among realities too, therefore, there are degrees of freedom. There are gaps to be exploited or filled, and work to be done to maintain and expand. The shifting ground between saving and neoliberal, corporate seed ordering is the most obvious example present in this book, but there are others suggestive of the complicated interactions of seed-people realities and the possibilities of changing them. For instance, seed and saver refigurations through encoding intellectual property rights (order a) facilitate private capital ownership, profit-making and consolidation (order b), but elaborations of farmers' rights (order c) contest this collaboration indicating ways in which relations might be altered to benefit savers (and seeds). In addition, reconsidering how worlds connect and disconnect, genebanking might be reworked to better serve conservation and farmer realities, and genetic engineering need not be left to private capital agendas. Seed saving may be threatened, but it was not always, and need not remain, this way.

Forging Ahead

A focus on practice and relations facilitates a more complex understanding of problems and points to ways in which worlds might be enacted differently, but it risks losing a sense of practitioners and their experiences. Paying attention to how and why particular humans and nonhumans engage in the ways they do offers key insights into how change might be fostered. Connecting everyday experiences and practices to broader concerns is part of building this understanding, and the lives and articulations of those involved should not be ignored. We need the attachments that bring material, sensory, cognitive engagements in order to keep open possibilities raised in seed saving, and to prise open those gaps within other worlds to support flourishing of seeds and people, and sustainable agrifood futures. It is, in other words, both the practitioners and the practices that matter.

In recognising that multiple participants come together to enact worlds, decisions about who/what are responsible may become murky. The solution, however, is not to return to less complex approaches but to appreciate that we are all – humans and nonhumans – implicated, to varying degrees. Not everyone will save seed, but there are diverse ways of enabling such practices. Our shared task is not only to be open to and learn to discern difference with nonhumans, but to choose and enact better worlds together. To improve our shared worlds requires rethinking and reworking existing practices as well as fostering new possibilities. This will not be simple or without conflict. We will need both creativity and critique.

Despite increasing acknowledgement that environmental and food challenges are beyond business-as-usual solutions, the corporate seed order is advancing, justifying its expansions as solutions. In this book I have tried to illustrate ways in which seed saving contributes to different relations and opportunities than those offered by ag-biotech and its advocates. I have intended also to open space for conversation about what a seed (and agrifood) order should achieve, as well as what can be done to support such an order. Seed saving, I have suggested, saves more than seeds – it saves possibilities. Whether and how these opportunities are taken up while they exist may well prove fundamental in reconstituting more responsible and sustainable agricultural and environmental relations.

Appendix 1

Summary Characterisation of Seed Saver Interviewees, by Pseudonym

Pseudonym	Region	Type of Space	Years Saving[1]	Number of Varieties[2]	Type of Activity[3]
Ada	Western	Garden	6–15	11–30	Leisure
Adam	Western	Farm	6–15	Over 100	Full-time
Alice	Eastern	Gardens	6–15	31–50	Leisure
Andrew	Central	Farm	16–30	11–30	Leisure
Anna	Western	Farm	6–15	Many	Leisure
Bella	Eastern	Garden	16–30	11–30	Leisure
Beth	Eastern	Garden	1–5	11–30	Leisure
Brent	Eastern	Farm	6–15	Over 100	Part-time
Carl	Western	Farm	6–15	11–30	Part-time
Carolyn	Central	Farm	6–15	11–30	Part-time
Conrad	Eastern	Garden	6–15	1–10	Leisure

1   Participants have been placed within the categories of saving seed for 1–5, 6–15, 16–30 or over 30 years. Estimations by participants were often vague, such as '10 years or so', and several mentioned childhood experiences but more sustained engagement as adults. The categorisations here attempt a best fit with savers' comments.

2   Most interviewees offered estimates of the number of varieties they were saving or provided a sample list, but several indicated saving 'many' varieties instead. The category 'many' has been added to those of 1–10, 11–30, 31–50, 51–100 and over 100, which reflect the clustered responses received.

3   Type of activity has been noted as: leisure, part-time work or full-time work. This category cannot be used to estimate the number of hours invested, or to suggest level of commitment. Instead, it indicates whether saving activities contributed to a saver's primary income and to what degree.

Pseudonym	Region	Type of Space	Years Saving	Number of Varieties	Type of Activity
Dara	Western	Garden	6–15	11–30	Leisure
Denis	Central	Farm and Garden	16–30	Many	Leisure, Part-time
Gail	Western	Farm	16–30	many	Leisure
Hannah	Eastern	Garden	16–30	1–10	Leisure
Irene	Eastern	Farm	6–15	31–50	Part-time
Jane	Central	Farm	16–30	Over 100	Full-time
Lee	Eastern	Garden	Over 30	1–10	Leisure
Lena	Western	Garden	16–30	Many	Leisure
Martin	Central	Gardens	Over 30	Over 100	Full-time
Mary	Western	Gardens	1–5	Many	Leisure
Matthew	Western	Gardens	16–30	Many	Leisure, Part-time
Melanie	Eastern	Garden	6–15	11–30	Leisure
Monica	Eastern	Gardens	16–30	11–30	Leisure
Natalie	Western	Gardens	16–30	Over 100	Leisure
Pat	Eastern	Garden	Over 30	Many	Leisure
Paul	Western	Gardens	Over 30	51–100	Leisure
Rachel	Western	Garden	6–15	11–30	Leisure
Richard	Central	Farm	6–15	51–100	Part-time
Sandra	Eastern	Garden	16–30	11–30	Leisure

Pseudonym	Region	Type of Space	Years Saving	Number of Varieties	Type of Activity
Sarah	Western	Farm	16–30	Over 100	Full-time
Susan	Eastern	Garden	6–15	11–30	Leisure
Tom	Western	Garden	6–15	1–10	Leisure
Will	Western	Gardens	Over 30	Many	Leisure

Appendix 2
Sociodemographic Characteristics of Survey Respondents

Variable	Survey Sample	
Gender	Response Count	Response Per Cent
Female	261	76.3
Male	81	23.7
Age	Response Count	Response Per Cent
18–25	11	3.2
26–35	85	24.9
36–45	73	21.3
46–55	81	23.7
56–65	65	19.0
66–75	25	7.3
Over 75	2	0.6
Employment	Response Count	Response Per Cent
Employed (full-time)	118	34.5
Employed (part-time)	41	12.0
Self-employed	84	24.6
Unemployed	10	2.9
Home duties	18	5.3
Retired	41	12.0
Student (full-time)	10	2.9
Student (part-time)	1	0.3
Other	19	5.6

Variable	Survey Sample	
Education	Response Count	Response Per Cent
Primary school	1	0.3
High school	24	7.0
CEGEP	7	2.0
Diploma	18	5.3
College	63	18.4
Trade/Apprenticeship	16	4.7
Bachelor degree	123	36.0
Master degree	64	18.7
Doctoral degree	10	2.9
Other	16	4.7
Years Saving Seed	Response Count	Response Per Cent
1–5	187	54.7
6–10	64	18.7
11–20	51	14.9
21–30	24	7.0
31–40	12	3.5
Over 40	4	1.2
Growing Space Type	Response Count (n=661)[1]	Response Per Cent
Backyard garden	229	67.0
Frontyard garden	124	36.3
Balcony/Patio	65	19.0
Community garden	74	21.6
Hobby farm	54	15.8
Farm	65	19.0
Other	50	14.6
Dwelling Type	Response Count	Response Per Cent
Detached house	259	75.7
Semi-detached house	35	10.2
Camper or cabin	8	2.3
Apartment or unit	32	9.4
Other	8	2.3

1 Participants indicated all types of growing spaces that applied.

Variable	Survey Sample	
Province/Territory	Response Count (n=333)	Response Per Cent
BC	82	24.6
AB	19	5.7
SK	14	4.2
MB	7	2.1
ON	123	36.9
QC	45	13.5
NB	10	3.0
NS	23	6.9
PEI	2	0.6
NL	6	1.8
YT	2	0.6
NT	0	0
NU	0	0
Household Composition	Response Count (n=339)	Response Per Cent
1 Adult	58	17.1
2 Adults	228	67.3
3 Adults	35	10.3
4 Adults	13	3.8
5 Adults	5	1.5
0 Children	246	71.9
1 Child	30	8.8
2 Children	46	13.5
3 Children	18	5.3
4 Children	2	0.6
Total Annual Household Income (After Tax)	Response Count	Response Per Cent
under 10,000	23	6.7
10–14,999	12	3.5
15–19,999	16	4.7
20–24,999	22	6.4
25–34,999	44	12.9
35–49,999	65	19.0
50–74,999	80	23.4
75–99,999	35	10.2
100–149,999	36	10.5
150,000 and over	9	2.7

Appendix 3

Food and Forage Crops Listed in Annex I of the ITPGRFA

Food Crops Listed in Annex I of the ITPGRFA

Crop	Genus	Crop	Genus
Apple	Malus	Lentil	Lens
Asparagus	Asparagus	Maize	Zea*
Banana/Plantain	Musa*	Major aroids	Colocasia, Xanthosoma*
Barley	Hordeum	Oat	Avena
Beans	Phaseolus*	Pea	Pisum
Beet	Beta	Pearl Millet	Pennisetum
Brassica complex	Brassica et al.*	Pigeon Pea	Cajanus
Breadfruit	Artocarpus*	Potato	Solanum*
Carrot	Daucus	Rice	Oryza
Cassava	Manihot*	Rye	Secale
Chickpea	Cicer	Sorghum	Sorghum
Citrus	Citrus*	Strawberry	Fragaria
Coconut	Cocos	Sunflower	Helianthus
Cowpea et al.	Vigia	Sweet Potato	Ipomoea
Eggplant	Solanum*	Triticale	Triticoseclae
Faba Bean/Vetch	Vicia	Wheat	Triticum et al.*
Finger Millet	Eleusine	Yams	Dioscorea
Grass pea	Lathyrus		

*Caveats apply

Fodder Crops Listed in Annex I of the ITPGRFA

Genera	Species
Agropyron	cristatum, desertorum
Agrostis	stolonifera, tenuis
Alopecurus	pratensis
Andropogon	gayanus
Arrhenatherum	elatius
Astragalus	chinensis, cicer, arenarius
Atriplex	halimus, nummularia
Canavalia	ensiformis
Coronilla	varia
Dactylis	glomerata
Festuca	arundinacea, gigantea, heterophylla, ovina, pratensis, rubra
Hedysarum	coronarium
Lathyrus	cicera, ciliolatus, hirsutus, ochrus, odoratus, sativus
Lespedeza	cuneata, striata, stipulacea
Lolium	hybridum, multiflorum, perenne, rigidum, temulentum
Lotus	corniculatus, subbiflorus, uliginosus
Lupinus	albus, angustifolius, luteus
Medicago	arborea, falcata, sativa, scutellata, rigidula, truncatula
Melilotus	albus, officinalis
Onobrychis	viciifolia
Ornithopus	sativus
Phalaris	aquatica, arundinacea
Phleum	pretense
Poa	alpina, annua, pratensis
Prosopis	affinis, alba, chilensis, nigra, pallida
Pueraria	phaseoloides
Salsola	vermiculata
Trifolium	alexandrinum, alpestre, ambiguum, angustifolium, arvense, agrocicerum, hybridum, incarnatum, pratense, repens, resupinatum, rueppellianum, semipilosum, subterraneum, vesiculosum
Tripsacum	laxum

Adapted from: FAO 2009

References

AAFC (Agriculture and Agri-Food Canada). nd. *Public Consultation Document: Frequently Asked Questions*. Working Group on Low Level Presence. AGRIDOC #2821497. Ottawa: Agriculture and Agri-Food Canada.

Abergel, E. 2000. *Growing Uncertainty: The Environmental Risk Assessment of Genetically Engineered Herbicide Tolerant Canola in Canada*. PhD diss., York University, Toronto.

Abergel, E. and Barrett, K. 2002. Putting the cart before the horse: a review of biotechnology policy in Canada. *Journal of Canadian Studies* 37(3), 135–61.

Adolphe, D. 2011. *Seed Certification in Canada*. Powerpoint presentation. Ottawa: Canadian Seed Growers' Association. [Online]. Available at http://www.seedforthefuture.ca/uploads/userfiles/files/What%20is%20 Certified%20Seed.pdf [accessed: 7 Dec. 2012].

Agreement on Trade-Related Aspects of Intellectual Property Rights, 1994. Annex IC, Final Act Embodying the Results of the Uruguay Round of Multilateral Trade Negotiations. Geneva: World Trade Organization.

Ahmed, S. 2010. Orientations matter, in New *Materialisms: Ontology, Agency, and Politics*, edited by D. Coole and S. Frost. Durham: Duke University Press, 234–57.

Alkon, A. and Mares, T. 2012. Food sovereignty in US food movements: radical visions and neoliberal constraints. *Agriculture and Human Values*, 29(3), 347–59.

Allen, D. Vice-President, Canadian Wheat Board, personal communication, Aug. 2007.

Allen, P. 2004. *Together at the Table: Sustainability and the Sustenance in the American Agri-food System*. University Park: Pennsylvania State University Press.

———. 2012. Divergence and convergence in alternative agrifood movements: seeking a path forward. Paper presented at *XIII World Congress of Rural Sociology*, Lisbon, Portugal, 31 July–4 Aug.

Allen, P. and Guthman, J. 2006. From 'old school' to 'farm-to-school': neoliberalization from the ground up. *Agriculture and Human Values*, 23(4), 401–15.

Altieri, M. 2005. The myth of coexistence: why transgenic crops are not compatible with agroecologically based systems of production. *Bulletin of Science, Technology and Society*, 25(4), 361–71.

Altieri, M. and Rosset, P. 2002. Ten reasons why biotechnology will not ensure food security, protect the environment and reduce poverty in the developing world, in *Ethical Issues in Biotechnology*, edited by R. Sherlock and J. Morrey. Oxford: Rowman and Lifflefield, 175–82.

Amagassa, K. 2007. *Spilled GM Canola Growing in Japan: Citizens' Survey Results 2007*. [Online] Available at: http://www.biosafety-info.net/bioart.php?bid=473&ac=st [accessed: 7 Dec. 2012].

Anand, N. 2006. Planning networks: processing India's national biodiversity strategy and action plan. *Conservation and Society*, 4(3), 471–87.

Anderson, B. and Harrison, P., eds. 2010. *Taking Place: Non-Representational Theories and Geography*. Surrey: Ashgate.

Anderson, B. and Wylie, J. 2009. On geography and materiality. *Environment and Planning A*, 41(2), 318–35

Anderson, K. 1997. A walk on the wild side: a critical geography of domestication. *Progress in Human Geography*, 21(4), 463–85.

Andersen, R. 2005. *The History of Farmers' Rights: A Guide to Central Documents and Literature*. Background Study 1, FNI Report 8. Lysaker: Fridtjof Nansens Institut.

———. 2008. *Governing Agrobiodiversity: Plant Genetics and Developing Countries*. Aldershot: Ashgate.

———. 2011. *The 2010 Global Consultation on Farmers' Rights*. FNI Report 2/2011. Lysaker: Fridtjof Nansen Institute.

Andrée, P. 2002. The biopolitics of genetically modified organisms in Canada. *Journal of Canadian Studies*, 37(3), 162–91.

———. 2007. *Genetically Modified Diplomacy: The Global Politics of Agricultural Biotechnology and the Environment*. Vancouver: UBC Press.

Appadurai, A. 1986. Introduction: commodities and the politics of value, in *The Social Life of Things: Commodities in Cultural Perspective*, edited by A. Appadurai. Cambridge: Cambridge University Press, 3–23.

Attenborough, D. 1998. *The Limits of Endurance: The Life of Birds*. [dir. J. Sarsby, BBC].

Ausubel, K. 1994. *Seeds of Change: The Living Treasure*. New York: Harper Collins.

Bakker, K. 2009. Neoliberal nature, ecological fixes, and the pitfalls of comparative research. *Environment and Planning A*, 41(8), 1781–7.

———. 2010. The limits of 'neoliberal natures': debating green neoliberalism. *Progress in Human Geography*, 34(6), 715–35.

Bakker, K. and Bridge, G. 2006. Material worlds? resource geographies and the 'matter of nature'. *Progress in Human Geography*, 30(1), 5–27.

Ball, G. 2012. *Burpee, GMO and Monsanto Rumors Put to Rest*. [Online] Available at: http://www.burpee.com/gygg/content.jsp?contentId=about-burpee-seeds [accessed: 7 Dec. 2012].

Bamford, S. 2002. On being 'natural' in the rainforest marketplace: science, capitalism and the commodification of biodiversity. *Social Analysis*, 46(1), 35–50.

Barad, K. 2007. *Meeting the Universe Halfway: Quantam Physics and the Entanglement of Matter and Meaning*. Durham: Duke University Press.

Barazani, O., Perevolotsy, A. and Hadas, R. 2007. A problem of the rich: prioritizing local plant genetic resources for *ex situ* conservation in Israel. *Biological Conservation*, 141(2), 596–600.

BASF. 2007. *2007 Season Clearfield Commitment for Canola*. BASF SE.

BEDE (Biodiversity Exchange and Diffusion of Experience) and RSP (Réseau Semences Paysannes). 2011. *Seeds and Farmers' Rights: How International Regulations Affect Farmer Seeds*. Montpellier and Brens: BEDE/RSP.

Beisel, U. 2010. Jumping hurdles with mosquitoes? *Environment and Planning D*, 28(1), 46–9.

Beingessner, P. 2004a. Canada: Seed Sector Review report pushes for changes. *Cropchoice.com*. [Online] Available at: http://www.cropchoice.com/leadstry25fa.html?recid=2561 [accessed: 7 Dec. 2012].

———. 2004b. Farmers beware: Canadian government getting out of the plant breeding business. *The New Farm*. [Online]. Available at: http://newfarm.rodaleinstitute.org/columns/saskatchewan/2004/1204/121704.shtml. [accessed: 7 Dec. 2012].

Belasco, W. and Horowitz, R. 2009. *Food Chains: From Farmyard to Shopping Cart*. Philadelphia: University of Pennsylvania Press.

Bellon, M.R. 2003. Conceptualizing interventions to support on-farm genetic resource conservation. *World Development*, 32(1), 159–72.

Bellon, M.R. and Berthaud, J. 2006. Traditional Mexican agriculture systems and the potential impacts of transgenic varieties on maize diversity. *Agriculture and Human Values*, 23, 3–14.

Benbrook, C. 1999. *Evidence of the Magnitude and Consequences of the Roundup and Consequences of the RoundUp Ready Soybean Yield Drag from University-Based Varietal Trials in 1998*. AgBiotech InfoNet Technical Paper 1. Sandpoint: Benbrook Consulting Services.

———. 2002. *Premium Paid for Bt Corn Seed Improves Corporate Finances while Eroding Grower Profits*. Sandpoint: Benbrook Consulting Services.

———. 2012. Impacts of genetically modified crops on pesticide use: the first sixteen years. *Environmental Sciences Europe*, 24, 24.

Bennett, E. 1968. FAO/IBP Technical Conference on the Exploration, Utilization and Conservation of Plant Genetic Resources. Rome: FAO.

Bennett, J. 2001. *The Enchantment of Modern Life: Attachments, Crossings, and Ethics*. Princeton: Princeton University Press.

———. 2010. *Vibrant Matter: A Political Ecology of Things*. Durham: Duke University Press.

Berry, W. 2002. The whole horse: the preservation of the agrarian mind, in *The Fatal Harvest Reader*, edited by A. Kimbell. Sausalito: Foundation for Deep Ecology, 39–48.

Bijker, W. and Law, J., eds. 1997. *Shaping Technology/Building Society: Studies in Sociotechnical Change,* second edition. Cambridge: MIT Press.

Bingham, N. 2006. Bees, butterflies, and bacteria. *Environment and Planning A*, 38, 483–98.

Bingham, N. and Hinchliffe, S. 2008. Reconstituting natures: articulating other modes of living together. *Geoforum*, 39(1), 83–7.

Biodiversity International. 2007. *Guidelines for the Development of Crop Descriptor Lists*. Biodiversity Technical Bulletin Series. Rome: Biodiversity International.

Bissell, D. 2008. Comfortable bodies: sedentary affects. *Environment and Planning A*, 40(7), 1697–712.

Blacksheep Strategy Inc. 2004. *Assessing Certified Seed Market Opportunities: Summary Report*. Report to the Canadian Plant Technology Agency. Winnipeg.

Blakeney, M. 2009. *Intellectual Property Rights and Food Security*. Oxfordshire: CABI.

Bocking, S. 2000. Agendas, interests and authority: science and politics in Canada. *Journal of Canadian Studies*, 37(3), 5–13.

Boehm, T. 2006. Sterile Seeds. *Deconstructing Dinner*. Radio communication, 9 Feb, at CJLY Kootenay Co-op Radio, Nelson, BC.

Borgmann, A. 1984. *Technology and the Character of Contemporary Life: A Philosophical Inquiry*. Chicago: University of Chicago Press.

———. 1995. The moral significance of the material culture, in *Technology and the Politics of Knowledge*, edited by A. Feenberg and A. Hannay. Indianapolis: Indiana University Press, 85–93.

Borowiak, C. 2004. Farmers' rights: intellectual property regimes and the struggle over seeds. *Politics and Society*, 32(4), 511–43.

Bourdieu, P. 1977. *Outline of a Theory of Practice*. Cambridge: Cambridge University Press.

Braun, B. and Whatmore, S., eds. 2010. *Political Matter: Technoscience, Democracy, and Public Life*. Minneapolis: University of Minnesota Press.

Brockway, L. 1988. Plant science and colonial expansion: the botanical chess game, in *Seeds and Sovereignty: The Use and Control of Plant Genetic Resources*, edited by J. Kloppenburg. Durham, NC: Duke University Press, 49–66.

———. 2002. *Science and Colonial Expansion: The Role of the British Royal Botanic Gardens*. New Haven: Yale University Press.

Brookes, G. and Barfoot, P. 2006. *GM Crops: The First Ten Years – Global Socio-Economic and Environmental Impacts*. ISAAA Brief 36. Ithaca: International Service for the Acquisition of Agri-Biotech Applications.

————. 2011. *GM Crops: Global Socio-Economic and Environmental Benefits 1996–2009*. Dorchester: PG Economics.

Brown, A.H.D., Frankel, O.H., Marshall, D.R. and Williams, J.T., eds. 1989. *The Use of Plant Genetic Resources*. Cambridge: Cambridge University Press.

Bruer, R. 2012. EU May Ease Genetic Engineering Rules on Food. *Deutsche Welle*. [Online]. Available at: http://www.dw.de/eu-may-ease-genetic-engineering-rules-on-food/a-16017969 [accessed: 7 Dec. 2012].

Brush, S., ed. 2000. *Genes in the Field: On-Farm Conservation of Crop Diversity*. Ottawa: IDRC.

————. 2004. *Farmer's Bounty: Locating Crop Diversity in the Contemporary World*. New Haven: Yale University Press.

————. 2007. Farmers' rights and protection of traditional agricultural knowledge. *World Development*, 35(9), 1499–514.

Buttel, F. 2005. The environmental and post-environmental politics of genetically modified crops and foods. *Environmental Politics*, 14(3), 309–23.

Bullard, L. 2005. *Freeing the Free Tree: A Briefing Paper on the First Legal Defeat of a BioPiracy Patent: The Neem Case*. New Delhi: Research Foundation for Science, Technology and Ecology.

Callicott, J.B. 1999. *Beyond the Land Ethic: More Essays in Environmental Philosophy*. Albany: State University of New York Press.

Callon, M. 1991. Techono-economic networks and irreversibility, in *A Sociological of Monsters: Essays on Power, Technology and Domination*, edited by J. Law. London: Routledge, 132–61.

Carew, R. 2000. Intellectual property rights: implications for the canola sector and publicly funded research. *Canadian Journal of Agricultural Economics*, 48, 175–94.

Carolan, M. 2010. *Decentring Biotechnology*. Surrey: Ashgate.

————. 2011. *Embodied Food Politics*. Surrey: Ashgate.

Castree, N. 2010a. Neoliberalism and the biophysical world 1: what 'neoliberalism' is and what difference nature makes to it. *Geography Compass*, 4(12), 1725–33.

————. 2010b. Neoliberalism and the biophysical world 2: theorising the neoliberalisation of nature. *Geography Compass*, 4(12), 1734–46.

————. 2011. Neoliberalism and the biophysical world 3: from theory to practice. *Geography Compass*, 5(1), 35–49.

CBAC (Canadian Biotechnology Advisory Committee). 2002. Patenting on Higher Life Forms: Report to the Government of Canada, Biotechnology Ministerial Coordinating Committee. Ottawa: CBAC.

CBAN (Canadian Biodiversity Action Network). 2006a. *The Battle in Brazil: Lines Drawn in Terminator Seed Fight*. [Online] Available at: http://banterminator.org/News-Updates/Archived-COP8-Press-Kit/1.-Media-Advisory. [accessed: 7 Dec. 2012].

———. 2006b. *Terminator Seeds: Canadian Government Admits it has No Capacity to Assess Impact on Farmers.* [Online]. Available at: http://www. cban.ca/Press/Press-Releases/Terminator-Seeds-Canadian-Government-admits-no-capacity-to-assess-impact-on-farmers. [accessed: 7 Dec. 2012].

———. 2012. *Flax.* CBAN. [Online]. Available at: http://www.cban.ca/ Resources/Topics/GE-Crops-and-Foods-Not-on-the-Market/Flax. [accessed: 7 Dec. 2012].

CGC (Canadian Grain Commission). 2008. *Canadian Grain Exports. Crop year 2007–2008.* [Online]. Available at: http://www.grainscanada.gc.ca/ statistics-statistiques/cge-ecg/cgea-aecg-eng.htm [accessed: 7 Dec. 2012].

———. 2009. *Canadian Grain Exports. Crop year 2008–2009.* [Online]. Available at: http://www.grainscanada.gc.ca/statistics-statistiques/cge-ecg/ cgea-aecg-eng.htm [accessed: 7 Dec. 2012].

———. 2011. *Canadian Grain Exports. Crop year 2010–2011.* [Online]. Available at: http://www.grainscanada.gc.ca/statistics-statistiques/cge-ecg/ cgea-aecg-eng.htm [accessed: 7 Dec. 2012].

CFIA (Canadian Food Inspection Agency). 2002. *10-Year Review of Canada's Plant Breeders' Rights Act.* Ottawa, CFIA.

———. 2006a. *Preliminary Proposal: Seed Regulatory Framework Modernization Including Adjustments to the Variety Registration System.* Ottawa: CFIA.

———. 2006b. *Proposed Amendments to the Plant Breeders' Rights Act to Bring Existing Legislation into Conformity with the 1991 UPOV Convention.* Ottawa: CFIA.

———. 2007. *Overview and Update of the Seed Program Modernization Initiative.* Ottawa: CFIA.

CFS (Center for Food Safety). 2004. *Monsanto vs. US Farmers.* Washington, DC: CFS.

———. 2007. *Monsanto vs. US Farmers: Nov 2007 Update.* Washington, DC: CFS.

———. 2011. *Legal Memorandum.* [Online]. Available at: http:// www.centerforfoodsafety.org/wp-content/uploads/2011/04/Legal-Memorandum.FINAL_.pdf [accessed: 9 Oct. 2012].

CGIAR (Consultative Group on International Agricultural Research). 2008. *Controversial Court Patent Case for Simple Yellow Legume has Become Rallying Point for 'Biopiracy' Concerns.* International Centre for Tropical Agriculture. 2 May 2008. [Online]. Available at: http://webapp.ciat.cgiar. org/newsroom/release_31.htm [accessed: 7 Dec. 2012].

———. 2012. *Who We Are.* [Online]. Available at: www.cgiar.org/who-we-are [accessed: 7 Dec. 2012].

Christie, M. 2005. The cultural geography of gardens. *Geographical Review,* 94(3), iii–iv.

Clapp, J. 2005. The political economy of food aid in an era of agricultural biotechnology. *Global Governance*, 11, 467–85.

———. 2008. Illegal GMO releases and corporate responsibility: questioning the effectiveness of voluntary measures. *Ecological Economics*, 66(2–3), 348–58.

Clapp, J. and Fuchs, D., eds. 2009. *Corporate Power in Agrifood Governance*. Cambridge: MIT Press.

Clark, A. 2003. Regulation of GM Crops in Canada: Science-Based or …? Paper to the *Safe Food Conference*, Salt Spring Island, Canada. [Online]. Available at: www.plant.uoguelph.ca/research/homepages/eclark/pdf/science.pdf [accessed: 7 Dec. 2012].

———. 2005. *Brief Respecting Bill C-27, the Canadian Food Inspection Agency Enforcement Act*. Presented to a Standing Committee on Agriculture, Bill C-27. Ottawa, Canada. Apr. [Online]. Available at: http://www.plant.uoguelph.ca/research/homepages/eclark/pdf/CFIA_revised.pdf [accessed: 7 Dec. 2012].

Clark, N. 2002. The demon-seed: bioinvasion as the unsettling of environmental cosmopolitanism. *Theory, Culture & Society*, 19(1–2), 101–25.

———. 2011. *Inhuman Nature: Sociable Life on a Dynamic Planet*. London: Sage.

COCERAL/FEDIOL. 2010. *Economic Impact Assessment: Low Level Presence of GMOs Not Authorized in Europe: The Linseed CDC Triffid case*. Brussels: COCERAL/FEDIOL.

COG (Canadian Organic Growers). 2005. *A Response to the Proposed Amendments to Plant Breeders' Rights Legislation and the Seed Sector Review*. Ottawa: Canadian Organic Growers.

Collins, H. 2006. Sterile Seeds. *Deconstructing Dinner*. Radio communication, 9 Feb, at CJLY Kootenay Co-op Radio, Nelson, BC.

Collins, H. and Krueger, R. 2003. *Potential Impact of GURTs on Smallholder Farmers, Indigenous & Local Communities and Farmers' Rights: The Benefits of GURTs*. Position paper of the International Seed Federation, submitted to the Convention on Biological Diversity's Ad Hoc Technical Expert Group on the Impact of GURTs on Smallholder Farmers, Indigenous People and Local Communities, 19–21 Feb.

Context Network. 2010. *Global Seed Market Values Grow by 10% to Reach Nearly $32B*. Press Release. Des Moines: Context Network. 10 May 2010. [Online]. Available at: http://www.seedquest.com/~seedqu5/news.php?type=news&id_article=6752&id_region=12&id_category=335&id_crop= [accessed: 7 Dec. 2012]

———. 2011. *Global Seed Market Value at $37B*. Press Release. 5 Jul. 2011. [Online]. Available at: http://www.seedquest.com/news.php?type=news&id_article=18823&id_region=&id_category=&id_crop= [accessed: 27 Oct. 2012].

Convention on Biological Diversity, 1992. Rome: United Nations.

Conway, G. 2000. Genetically modified crops: risks and promise. *Conservation Ecology*, 4(1), 2.

Cook, I. et al. 2006. Geographies of food: following. *Progress in Human Geography*, 30(5), 655–66.

Cook, I., Jackson, P., Hayes-Conroy, A., Abrahamsson, S., Sandover, R., Sheller, M., Henderson, H., Hallett IV, L., Imai, S., Maye, D. 2013.. Food's cultural geographies: texture, creativity and publics, in *The Wiley-Blackwell Companion to Cultural Geography*, edited by N. Johnson, R. Schein, J. Winders. Oxford: Wiley-Blackwell.

Coole, D. 2005. Rethinking agency: a phenomenological approach to embodiment and agentic capacities. *Political Studies*, 53, 124–42.

Coole, D. and Frost, S., eds. 2010. *New Materialisms: Ontology, Agency, and Politics.* Durham: Duke University Press.

Cornucopia Institute. 2012. *Prop 37: Your Right to Know.* [Online]. Available at: http://www.cornucopia.org/2012/09/california-proposition-37-your-right-to-know-what-is-in-your-food/ [accessed: 7 Nov. 2012].

Correa, C. 2000. *Options for the Implementation of Farmers' Rights at the National Level.* Working Paper 8. Geneva: South Centre.

Crang, P. 2010. Cultural geography: after a fashion. *Cultural Geographies*, 17(2), 191–201.

CropLife International. 2012a. *Low Level Presence.* [Online]. Available at: http://www.croplife.org/low_level_presence [accessed: 7 Dec. 2012].

———. 2012b. *Co-existence: Biotech, Conventional, and Organic.* [Online]. Available at: http://www.croplife.org/coexistence [accessed: 7 Dec. 2012].

Crouch, D. 2003a. Performances and constitutions of natures: a consideration of the performance of lay geography, in *Nature Performed: Environment, Culture and Performance,* edited by B. Szersynski, W. Heim, and C. Waterton. Oxford: Blackwell Publishing, 17–30.

———. 2003b. Spacing, performing, and becoming: tangles in the mundane. *Environment and Planning A*, 35, 1945–60.

Crucible Group II. 2000. *Seeding Solutions.* Ottawa, Rome, Upsaala: International Development Research Centre, International Plant Genetic Resource Institute, Dag Hammarskjöld Foundation.

CSA (Canadian Seed Alliance). 2004. *The Report of the Seed Sector Advisory Committee.* Ottawa: Canadian Seed Alliance. May 2004.

CSGA (Canadian Seed Growers Association). 2005. Certified Seed: Is the Message Getting Through? *Seed Scoop*, Apr. Ottawa: CSGA.

———. 2010. Tips for selling seed. *Seed to Succeed.* Ottawa: CSGA, 17–18.

———. 2011. *Annual Report 2010/11.* Ottawa: CSGA.

———. nd. *Top 10 Reasons to Use Certified Seed.* Ottawa: CSGA.

CSTA (Canadian Seed Trade Association). 2004a. *Fast Facts on the Canadian Seed Industry*. Mar. [Online]. Accessed at: http://www.cdnseed.org, no longer available.

———. 2004b. *Canadian Seed Trade Association Addresses Seed Saving Issue in Supreme Court Appeal*. News Release. 23 Jan. [Online]. Available at: http://cdnseed.org/archive/pdfs/press/CSTA%20Addresses%20Seed%20 Saving%20Issue.pdf [accessed: 7 Dec. 2012].

———. 2011a. *Canada's Seed Industry*. Powerpoint presentation. [Online]. Available at: cdnseed.org/wp-content/uploads/2011/08/About-the-Seed-Industry1.pdf [accessed: 4 Dec. 2012].

———. 2011b. *Fostering Innovation in Canadian Agriculture*. Submission to the Minister of Agriculture and Agri-Food Canada. Feb. [Online]. Available at: http://cdnseed.org/archive/pdfs/Presentations2011/MP submissionFebruary2011.pdf [accessed: 7 Dec. 2012].

———. nd. *The Value of Certified Seed*. Ottawa: CSTA. [Online]. Available at: http://cdnseed.org/wp-content/uploads/2011/08/The-Value-of-Certified-Seed.pdf [accessed: 7 Dec. 2012].

Cubby, B. 2011. GM wheat trial begins amid secrecy. *Sydney Morning Herald*. 28 May 2011. [Online]. Available at: http://www.smh.com.au/nsw/gm-wheat-trial-begins-amid-secrecy-20110527-1f8hl.html [accessed: 15 Nov. 2012].

Dawson, A. 2010a. CFIA says its getting out of seed certification. *Manitoba Cooperator*. 22 Jul. [Online]. Available at: http://www.agcanada.com/ manitobacooperator/2010/07/22/cfia-says-its-getting-out-of-seed-certification/ [accessed: 13 Oct. 2012].

———. 2010b. Voluntary better than legislated. *Manitoba Cooperator*. 29 Apr. [Online]. Available at: http://www.agcanada.com/ manitobacooperator/2010/04/29/voluntary-better-than-legislated/ [accessed: 13 Oct. 2012].

De Beer, J. 2005. Reconciling property rights in plants. *Journal of World Intellectual Property*, 8(1), 5–31.

———. 2007. The rights and responsibilities of biotech patent holders. *University of British Columbia Law Review*, 40(1), 343–73.

de Certeau, M. 1984. *The Practice of Everyday Life*, trans. S. Rendall. Los Angeles: University of California Press.

de Laet, M. 2000. Patents, travel, space: ethnographic encounters with objects in transit. *Environment and Planning D*, 18, 149–68.

de Laet, M. and Mol, A. 2000. The Zimbabwe bush pump: mechanics of a fluid technology. *Social Studies of Science*, 30(2), 225–63.

DeShutter, O. 2009. *Seed Policies and the Right to Food: Enhancing Agrobiodiversity and Encouraging Innovation*. Submission to the 64th session of the United Nations' General Assembly. A/64/170.

Declaration of Nyéléni. 2007. *Declaration of Forum for Food Sovereignty*, 23–7 Feb. 2007, Sélingué, Mali. [Online]. http://www.nyeleni.org/spip.php?article290 [accessed: 11 Oct. 2012].

Diederichsen, A. Curator, Plant Gene Resources Canada, personal communication, Mar. 2012.

Del Rio, A. and Bamberg, J. 2003. The effect of genebank seed increase on the genetics of recently collected potato (Solanum) germplasm. *American Journal of Potato Research*, 80(3), 215–18.

De Vincente, M.C. and Andersson, M.S., eds. 2006. *DNA Banks: Providing Novel Options for Genebanks?* Rome: International Plant Genetic Resources Institute.

Dillon, M. 2005. *'And We Have the Seeds': Monsanto Purchases World's Largest Vegetable Seed Company.* [Online]. Available at: http://www.seedalliance.org/seed_news/seminismonsanto/ [accessed: 2 Oct. 2012].

Dow AgroSciences. 2007. *Why Biotech is Important. Plant Genetics and Biotechnology.* [Online]. Accessed at: http://www.dowagro.com/pgb/intro/why, no longer available.

DPL (Delta and PineLand Company). 2005. *Technology Protection System: Providing the Potential to Enhance Biosafety and Biodiversity in Production Agriculture.* Pamphlet. Scott: Delta and Pine Land Company.

DPL (Delta and PineLand Company) and USA (United States of America). 1998. *Patent 5 723 765: Control of Plant Gene Expression.* United States Patent and Trade Office.

Dupont/Pioneer Hi-Bred. 2010. *Comments of DuPont/Pioneer Hi-Bred International Regarding Agriculture and Antitrust Enforcement Issues in Our 21st Century Economy.* Submitted to United States Department of Justice AntiTrust Division and United State Department of Justice. [Online]. Available at: http://www.justice.gov/atr/public/workshops/ag2010/comments/254990.pdf [accessed: 14 Nov. 2012].

Drayton, R. 2000. *Nature's Government: Science, Imperial Britain, and the "Improvement" of the World.* New Haven: Yale University Press.

Dupuis, E.M. and Goodman, D. 2005. Should we go "home" to eat? toward a reflexive politics of localism. *Journal of Rural Studies*, 21, 359–71.

Dutfield, G. 2008. Turning Plant Varieties into Intellectual Property: the UPOV Convention, in *The Future Control of Food*, edited by G. Tansey and T. Rajotte. London: Earthscan, 27–47.

———. 2009. *Intellectual Property Rights and the Life Science Industries*, second edition. Surrey: Ashgate.

———. 2011. *The Role of the International Union for the Protection of New Varieties of Plants (UPOV).* Paper 9. Geneva: Quaker United Nationals Office.

Dyck, B., Fries, T., Glen, B., McMillan, D. and Zacharias, K. 2010. Editorial. *Western Producer.* 14 Jan.

Eaton, D., Van Tongeren, F., Louwaars, N., Visser, B. and Van der Meer, I. 2002. Economic and policy aspects of "Terminator" technology. *Biotechnology and Development Monitor*, 49, 19–22.

Eden, S., Bear, C. and Walker, G. 2008. Mucky carrots and other proxies: problematising the knowledge-fix for sustainable and ethical consumption. *Geoforum*, 39(2), 1044–57.

Ellis, R. 2011. Jizz and the joy of pattern recognition: virtuosity, discipline and the agency of insight in UK naturalists' arts of seeing. *Social Studies of Science*, 41(6), 769–90.

Emerson, R.W. 1878. *Fortune of the Republic*. Boston: Houghton and Osgood.

Engdahl, F.W. 2006. Monsanto buys "Terminator" seeds company. *Global Research*. 27 Aug. [Online]. Availabe at: http://www.globalresearch. ca/monsanto-buys-terminator-seeds-company/3082 [accessed: 22 Oct. 2012].

Engels, J. 2001. Home gardens: a genetic resources perspective, in *Proceedings of the Second International Home Gardens Workshop: Contribution of Home Gardens to in situ Conservation of Plant Genetic Resources in Farming Systems*, edited by J.W. Watson and P.B. Eyzaguirre. Rome: International Plant Genetic Resources Institute, 3–9.

Engels, J. and Visser, B., eds. 2003. *A Guide to Effective Genebank Management. Biodiversity International*. Handbook 6. Rome: International Plant Genetic Resource Institute.

Escobar, A. 1998. Whose knowledge, whose nature? biodiversity, conservation and the political ecology of social movements. *Journal of Political Ecology*, 5(1), 53–82.

———. 2006. *Places and Regions in the Age of Globality: Social Movements and Biodiversity Conservation in the Colombian Pacific*. [Online]. Available at: http://www.unc.edu/~aescobar/text/eng/places_regions.doc [accessed: 3 Nov. 2012].

ETC Group. 2003. *Terminator Technology: Five Years Later*. Communiqué 79. Ottawa: ETC Group. [Online]. Available at: http://www.etcgroup. org/sites/www.etcgroup.org/files/publication/167/01/termcom03.pdf [accessed: 5 Oct. 2012].

———. 2004. *Canadian Supreme Court Tramples Farmers' Rights: Affirms Corporate Monopoly on Higher Life Forms*. Press Release, 21 May. [Online]. Available at: http://www.etcgroup.org/content/canadian-supreme-court-tramples-farmers-rights-affirms-corporate-monopoly-higher-life-forms [accessed: 5 Oct. 2012].

———. 2005. *Oligopoly Inc. 2005: Concentration in Corporate Power*. Communiqué 91. Ottawa: ETC Group. [Online]. Available at: http://www. etcgroup.org/content/oligopoly-inc-concentration-corporate-power-2005 [accessed: 5 Oct. 2012].

———. 2007a. *Updated: The World's Top 10 Seed Companies – 2006*. Ottawa: ETC Group. [Online]. Available at: http://www.etcgroup.org/content/updated-worlds-top-10-seed-companies-2006 [accessed: 5 Oct. 2012].

———. 2007b. *Terminator: The Sequel*. Communiqué 95. Ottawa: ETC Group. [Online]. Available at: http://www.etcgroup.org/content/terminator-sequel [accessed: 7 Oct. 2012]

———. 2008. *Who Own's Nature?* Ottawa: ETC Group. [Online]. Available at: http://www.etcgroup.org/content/terminator-sequel [accessed: 5 Oct. 2012].

———. 2011. *Who Will Control the Green Economy?* Communiqué 107. Ottawa: ETC Group. [Online]. Available at: http://www.etcgroup.org/content/who-will-control-green-economy-0 [accessed: 5 Oct. 2012].

———. 2012. *The Greed Revolution*. Communiqué 108. Jan/Feb. Ottawa: ETC Group. [Online]. Available at: http://www.etcgroup.org/content/greed-revolution [accessed: 5 Oct. 2012].

Ewins, A. 2010. GM opponents use flax to back views. *Western Producer*. 18 Feb. [Online]. Available at: http://www.producer.com/2010/02/gm-opponents-use-flax-to-back-views/ [accessed: 17 Oct. 2012].

Falcon, W.P. and Fowler, C. 2002. Carving up the commons: emergence of a new international regime for germplasm development and transfer. *Food Policy*, 27, 197–222.

FAO (Food and Agriculture Organisation of the United Nations). 1996. *Global Plan of Action for the Conservation and Sustainable Utilization of Plant Genetic Resources for Food and Agriculture and the Leipzig Declaration*, adopted by the International Technical Conference on Plant Genetic Resources, Leipzig Germany, 17–23 Jun. Rome: FAO.

———. 1997. *The State of the World's Plant Genetic Resources for Food and Agriculture*. Rome: FAO.

———. 1998. *Report: Sustaining Agricultural Biodiversity and Agro-Ecosystem Functions*. International Technical Workshop. Rome: FAO.

———. 2003. *Trade Reforms and Food Security: Conceptualizing the Linkages*. Rome: FAO.

———. 2004a. *State of the Food and Agriculture 2003–4. Agricultural Biotechnology: Meeting the Needs of the Poor?* Rome: FAO.

———. 2004b. What is Agrobiodiversity? Factsheet, in *Building on Gender, Agrobiodiversity and Local Knowledge*. Rome: FAO.

———. 2006. *Report of the Governing Body of the International Treaty on Plant Genetic Resources for Food and Agriculture*. IT/GB-1/06/Report: First Session, Madrid, Spain, 12–16 Jun.

———. 2010a. *Agricultural Biotechnologies in Developing Countries*. Prepatory Document ABDC 10/9. FAO International Technical Conference, Guadalajara, Mexico, 1–4 Mar.

———. 2010b. *The Second Report on the State of the World's Plant Genetic Resources for Food and Agriculture*. Rome: Commission on Genetic Resources for Food and Agriculture.

———. 2011. *Resolution 6/2011: Implementation of Article 9, Farmers' Rights*, 4th session of the Governing Body, International Treaty on Plant Genetic Resources for Food and Agriculture, Bali, Indonesia, 14–18 Mar.

———. 2012. *Second Global Plan of Action for Plant Genetic Resources for Food and Agriculture*. Adopted by the FAO Council, Rome, Italy, 29 Nov.

FAO/IPGRI (Food and Agriculture Organization and International Plant Genetic Resources Institute). 1994. *Genebank Standards*. Rome: FAO.

———. 2011. *Draft Updated Genebank Standards: Minimum Standards for Conservation of Orthodox Seeds*. Circulated 6 Jan. 2011. Rome: FAO. [Online]. Available at: http://www.fao.org/docrep/meeting/022/MB179E.pdf [accessed: 21 Oct. 2012].

FAOStat (Food and Agriculture Organisation Statistics). 2009. *Top Production – Linseed – 2009*. [Online]. Available at: http://faostat.fao.org/site/339/default.aspx [accessed: 21 Oct. 2012].

Farmers' Rights. 2012. Lysaker: Fridtjof Nansen Institute. [Online]. Available at: http://www.farmersrights.org [accessed: 5 Dec. 2012].

Farnham, T.J. 2002. *The Concept of Biological Diversity: The Evolution of a Conservation Paradigm*. PhD diss., Yale University, New Haven.

FCC (Flax Council of Canada). 2010a. *Message to Producers: Certified Seed*. 20 Jan. [Online]. Accessed at: http://www.flaxcouncil.ca, no longer available.

———. 2010b. *Flax Council of Canada Announces Industry Stewardship Program for Farm Saved Planting Seed*. Press Release. 26 Mar. [Online]. Available at: http://www.flaxcouncil.ca/files/web/NEWS%20RELEASE%20-%20Flax%20Council%20of%20Canada%20Announces%20Industry%20Stewardship%20Program%20for%20Farm%20Saved%20Planting%20Seed%2003.12.10%20F.pdf [accessed: 2 Oct. 2012].

———. 2011. *Flax Council of Canada Urges Growers to Test Planting Seed*. Press Release. 24 Mar. [Online]. Available at: http://www.flaxcouncil.ca/files/web/Flax%20Council%20of%20Canada%20Urges%20Growers%20to%20Test%20Planting%20Seed%20-%20March%202011%20web%20rev.2.pdf [accessed: 2 Oct. 2012].

———. 2012. *Canadian Flax Industry Update*. Spring. Winnipeg: FCC. [Online]. Available at: http://www.flaxcouncil.ca/files/web/FlaxIndustryUpdate.pdf [accessed: 2 Nov. 2012].

Fedco Seeds. 2005. *Monsanto Buys Seminis*. Waterville: Fedco. [Online]. Available at: http://www.fedcoseeds.com/seeds/articles/monsanto.htm [accessed: 8 Oct. 2012].

Fehr, S. 2005. The path forward. *Germination*, 9(3), 4–8.

Feyissa, R. 2006a. *Farmers' Rights in Ethiopia: A Case Study*. Background Study 5, FNI Report 7. Lysaker: Fridtjof Nansens Institut.

———. Executive Director, Ethio-Organic Seed Action, personal communication, Jul. 2006b.

FOEI (Friends of the Earth International) 2011. *Who Benefits from GM Crops?* Amsterdam: FOEI. [Online]. Available at: http://www.foei.org/en/resources/publications/pdfs/2011/who-benefits-from-gm-crops-2011/view [accessed: 4 Dec. 2012].

Foucault, M. 1990. *The History of Sexuality: An Introduction*, trans. R. Hurley. New York: Vintage Books.

Fowler, C. 1994. *Unnatural Selection: Technology, Politics and Plant Selection*. Yverdon: Gordon and Breach.

———. 2004. An international perspective on trends and needs in public agricultural research, in *Seeds and Breeds for the 21st Century, Summit Proceedings*, edited by M. Sligh and L. Lauffer. Pittsboro: Rural Advancement Foundation International, 1–9.

———. 2008. *The Svalbard Global Seed Vault: Securing the Future of Agriculture*. Rome: Global Crop Diversity Trust. [Online]. Available at: http://www.croptrust.org/documents/Svalbard%20opening/New%20EMBARGOED-Global%20Crop%20Diversity%20Trust%20Svalbard%20Paper.pdf [accessed: 5 Dec. 2012].

Fowler, C. and Hodgkin, T. 2004. Plant Genetic Resources for Food and Agriculture: Assessing Global Availability. *Annual Reviews Environmental Resources*, 29, 143–79

Fowler, C. and Mooney, P. 1990. *Shattering: Food, Politics, and the Loss of Genetic Diversity*. Tucson: University of Arizona Press.

Fox Keller, E. 1985. *Reflections on Gender and Science*. New Haven: Yale University Press.

———. 2000. *The Century of the Gene*. Cambridge: Harvard University Press.

Franklin, U. 1999. *The Real World of Technology*, revised edition. Toronto: Anansi.

Freidberg, S. 2004. *French Beans and Food Scares: Culture and Commerce in an Anxious Age*. Oxford: Oxford University Press.

Friedmann, H. 1993. The political economy of food: a global crisis. *New Left Review*, 197, 29–57.

———. 2005. From colonialism to green capitalism: social movements and the emergence of food regimes, in *New Directions in the Sociology of Global Development. Research in Rural Sociology and Development*, 11, edited by F.H. Buttel and P. McMichael. Oxford: Elsevier, 229–67.

Friesen, L.F., Nelson, A.G. and Van Acker, R.C. 2003. Evidence of contamination of pedigreed canola (Brassica Napus) seedlots in Western Canada with genetically engineered herbicide resistance traits. *Agronomy Journal*, 95, 1342–7.

Gates Foundation. 2012. *Why the Foundation Funds Research in Crop Biotechnology*. [Online]. Available at: http://www.gatesfoundation.

org/agriculturaldevelopment/Pages/why-we-fund-research-in-crop-biotechnology.aspx [accessed: 7 Dec. 2012].

GCDT (Global Crop Diversity Trust) 2002. *Crop Diversity at Risk: The Case for Sustaining Crop Collections*. Report prepared with the Department of Agricultural Sciences, Imperial College of Wye. [Online]. Available at: www.croptrust.org/documents/WebPDF/wyereport.pdf [accessed: 21 Oct. 2012].

———. 2004. *Annual Report*. Rome: GCDT.

———. 2011. *Annual Report*. Rome: GCDT.

———. 2012a. *Svalbard Global Seed Vault*. [Online]. Available at: http://www.croptrust.org/content/svalbard-global-seed-vault [accessed: 21 Oct. 2012].

———. 2012b. *What We Do: Using*. [Online]. Available at: http://www.croptrust.org/content/using [accessed: 21 Oct. 2012].

Giard, L. 1998. Part II: Doing cooking, in *The Practice of Everyday Life: Living and Cooking*, trans. T. Tomasik, edited by M. de Certeau, L. Giard and P. Mayol. Minneapolis: University Minnesota Press, 149–248.

Gibson-Graham, J.K. 2006. *A Postcapitalist Politics*. Minneapolis: University of Minnesota Press.

———. 2008. Diverse economies: performative practices for 'other worlds'. *Progress in Human Geography*, 32(5), 613–32.

———. 2011. A feminist project of belonging for the Anthropocene. *Gender, Place and Culture*, 18(1), 1–21.

Gibson-Graham, J.K. and Roelvink, G. 2009. A postcapitalist politics of dwelling: ecological humanities and community economies in conversation. *Australian Humanities Review*, 46, 145–58

Gillam, C. 2010a. *Bayer Settles US Rice Contamination Case*. Reuters. 19 Oct. [Online]. Available at: http://www.reuters.com/article/2010/10/19/us-rice-bayer-idUSTRE69I4GK20101019 [accessed: 11 Oct. 2012].

———. 2010b. *Monsanto Sees 'Right Time' for GMO Wheat*. Reuters. 4 Nov. [Online]. Available at: http://www.reuters.com/article/2010/11/04/us-monsanto-wheat-gmo-idUSTRE6A34K220101104 [accessed: 11 Oct. 2012].

Glover, D. 2008. *Made by Monsanto: The Corporate Shaping of GM Crops as a Technology for the Poor*. Working Paper 11. Brighton: STEPs Centre.

Goeschl, T. and Swanson, T. 2003. The development impact of genetic use restriction technologies. *Environment and Development Economics*, 8(1), 149–65.

Gómez, O.J., Blair, M.W., Frankow-Lindberg, B.E. and Gullberg, U. 2005. Comparative study of common bean (*Phaseolus vulgaris* L.) landraces conserved *ex situ* in genebanks and *in situ* by farmers. *Genetic Resources and Crop Evolution*, 52(4), 371–80.

Goodman, D. and Dupuis, E.M. 2002. Knowing food and growing food: beyond the production-consumption debate in the sociology of agriculture. *Sociologia Ruralis*, 42(1), 5–23.

GRAIN. 2005a. India's New Seed Law. *Seedling*, Jul. [Online]. Available at: http://www.grain.org/article/entries/457-india-s-new-seed-bill [accessed: 3 Oct. 2012].

———. 2005b. Africa's Seed Laws: Red Carpet for Corporations. *Seedling,* Jul. [Online]. Available at: http://www.grain.org/fr/article/entries/540-africa-s-seeds-laws-red-carpet-for-corporations [accessed: 3 Oct. 2012].

———. 2007. *A Gene Bank in Tatters*. [Online]. Available at: http://www.grain. org/fr/article/entries/4203-a-genebank-in-tatters [accessed: 3 Oct. 2012].

———. 2008. Faults in the Vault: Not everyone is celebrating Svalbard. *Against the Grain*. Barcelona: GRAIN. [Online]. Available at: http://www.grain. org/article/entries/181-faults-in-the-vault-not-everyone-is-celebrating-svalbard [accessed: 3 Oct. 2012].

Greenhough, B. 2010. Vitalist geographies: life and the more-than-human, in *Taking Place: Non-Representational Theories and Geography*, edited by B. Anderson and P. Harrison. Surrye: Ashgate, 37–54.

———. 2012. Where species meet and mingle: endemic human-virus relations, embodied communication and more-than-human agency at the Common Cold Unit 1946–90. *Cultural Geographies*, 19(3), 281–301.

Greenhough, B. and Roe, E. 2010. From ethical principles to response-able practice. *Environment and Planning D*, 28(1), 43–5.

———. 2011. Ethics, space, and somatic sensibilities: comparing relationships between scientific researchers and their human and animal experimental subjects. *Environment and Planning D*, 29(1), 47–66.

Guthman, J. 2007. Voluntary food labels as neoliberal governance. *Antipode*, 39(3), 456–78.

Hall, L., Topinka, K., Huffman, J., Davis, L., Good, A. 2000. Pollen flow between herbicide-resistant Brassica Napus is the cause of multiple-resistant B. Napus volunteers. *Weed Science*, 48(6), 688–94.

Hall, M. 2010. *Plants as Persons: A Philosophical Botany*. Albany: State University of New York Press.

Hammond, E. 2011. *How US Sorghum Distributions Undermine the FAO Plant Treaty Mechanism*. Briefing Paper. Oslo: Development Fund.

Haraway, D. 1988. Situated knowledges: the science question in feminism and the privilege of partial perspective. *Feminist Studies*, 14(3), 575–99.

———. 1989. *Primate Visions*. New York: Routledge.

———. 1991. A cyborg manifesto: science, technology, and socialist-feminsim in the late twentieth century, in *Simians, Cyborgs, and Women: The Reinvention of Nature*. London: Routledge, 149–81.

————. 1992. The promises of monsters: a regenerative politics for inappropriate/d others, in *Cultural Studies*, edited by L. Grossberg, C. Nelson and P.A. Treichler, London: Routledge, 295–337.

————. 1997. *Modest_Witness@Second_Millennium.FemaleMan©_Meets_ OncoMouse™*. London: Routledge.

————. 2003. *The Companion Species Manifesto: Dogs, People, and Significant Otherness*. Chicago: Prickly Paradigm Press.

————. 2008. *When Species Meet*. Minneapolis: University of Minnesota.

Harlan, J. 1975. *Crops and Man*. Madison: American Society of Agronomy and the Crop Science Society of America.

Harris, A. and Beasley, D. 2011. Bayer agrees to pay $750 million to end lawsuits over gene-modified rice. *Bloomberg News*. 2 Jul. [Online]. Available at: http:// www.bloomberg.com/news/2011-07-01/bayer-to-pay-750-million-to-end- lawsuits-over-genetically-modified-rice.html [accessed: 12 Nov. 2012].

Hassanein, N. 2003. Practicing food democracy: a pragmatic politics of transformation. *Journal of Rural Studies*, 19, 77–86.

Head, L. and J. Atchison. 2009. Cultural ecology: emerging human-plant geographies. *Progress in Human Geography*, 33(2), 236–45.

————. 2012. *Ingrained: A Human Bio-Geography of Wheat*. Surrey: Ashgate.

Heads Up: PBR in Practice. 2000. *Germination*, 4(5), 4.

Heldke, L. 1992. Foodmaking as thoughtful practice, in *Cooking, Eating, Thinking: Transformative Philosophies of Food*, edited by D. Curtin and L. Heldke. Indianapolis: Indiana State University, 203–29.

Helfer, L. 2004. *Intellectual property rights in plant varieties*. Rome: FAO.

Henderson, E. 2000. Rebuilding local food systems from the grassroots up, in *Hungry for Profit: The Agribusiness Threat to Farmers, Food and the Environment*, edited by F. Magdoff, J. Bellamy Foster and F.H. Buttel. New York: Monthly Review Press, 175–88.

Heynen, N. and Robbins, P. 2005. The neoliberalization of nature: governance, privatization, enclosure and valuation. *Capitalism Nature Socialism*, 16(1), 5–8.

Hills, M., Hall, L., Arnison, P., Good, A. 2007. Genetic use restriction technologies (GURTs): strategies to impede transgene movement. *Trends in Plant Science*, 12(4), 177–83.

Higgs, E. 2003. *Nature by Design: People, Natural Process, and Ecological Restoration*. London: MIT Press.

Highmore, B. 2002. *Everyday Life and Cultural Theory: An Introduction*. London: Routledge.

Hinchliffe, S. 2003. 'Inhabiting': landscapes and natures, in *Handbook of Cultural Geography*, edited by K. Anderson, M. Domosh, S. Pile and N. Thrift. London: Sage, 207–26.

————. 2007. *Geographies of Nature: Societies, Environments, Ecologies.* London: Sage.

————. 2010. A non-representational approach to environmental issues, in *Taking Place: Non-Representational Theories and Geography*, edited by B. Anderson and P. Harrison. Farnham: Ashgate, 303–20.

Hinchliffe, S., Kearnes, M.B., Degen, M., and Whatmore, S. 2005. Urban wild things: a cosmopolitical experiment. *Environment and Planning D*, 23, 643–58.

Hinchliffe, S. and Whatmore, S. 2006. Living cities: towards a politics of conviviality. *Science as Culture*, 15(2), 123–38.

Hinrichs, C.C. 2007. Introduction: practice and place in remaking the food system, in *Remaking the North American Food System: Strategies for Sustainability*, edited by C.C. Hinrichs and T.A. Lyson. Lincoln: University of Nebraska Press, 1–18.

Hinrichs, C.C. 2012. The system's in crisis: new openings and occlusions for agrifood movements. Paper presented at *XIII World Congress of Rural Sociology*, Lisbon, Portugal, 31 July–4 Aug.

Hird, M. 2009. *The Origins of Sociable Life.* Basingstoke: Palgrame MacMillan.

Hitchings, R. 2003. People, plants, and performance: on actor network theory and the material pleasures of the private garden. *Social and Cultural Geography*, 4, 99–114.

————. 2006. Expertise and inability: cultured materials and the reason for some retreating laws in London. *Journal of Material Culture*, 11, 364–81.

————. 2007. Approaching life in the London garden centre: acquiring entities and providing products. *Environment and Planning A*, 39(2), 242–59.

Hitchings, R. and Jones, V. 2004. Living with plants and the exploration of botanical encounter within human geographic research practice. *Ethics, Place & Environment*, 7(1), 3–18.

Holm, R. Professor, Department of Plant Sciences and Director, Crop Development Centre, University of Saskatchewan, personal communication, Aug. 2007.

Howard, P. 2009. Visualizing consolidation in the global seed industry: 1996–2008. *Sustainability*, 1(4), 1266–87.

Industry Canada. 1993. *Federal Expenditures for Biotechnology, 1989–1992.* Biotechnology Directorate. Ottawa: Industry Canada.

————. 2007. *Canada's New Government: Mobilizing Science and Technology to Canada's Advantage.* Ottawa: Industry Canada.

Ingold, T. 2000. *Perception of the Environment: Essays in Livelihood, Dwelling and Skill.* London: Routledge.

————. 2010. *Bringing Things to Life: Creative Entanglements in a World of Materials.* NCRM Working Paper 15, ESRC National Centre for Research Methods.

International Convention for the Protection of New Varieties of Plants. 1978. Geneva: UPOV.

——— 1991. Geneva: UPOV.

International Treaty on Plant and Genetic Resources for Food and Agriculture, 2009. Rome: FAO.

Investment in brown bag monitoring pays off. 2006. *Germination*, Oct. [Online]. Available at: http://www.seedquest.com/News/from/northamerica/ newsfromcanada.htm [accessed: 29 Nov. 2012].

ISF (International Seed Federation). 2006. *World Seed Trade Statistics*. [Online]. Available at: http://www.worldseed.org/statistics.htm [accessed: 5 Dec. 2006].

———. 2012. *Estimated Value of the Domestic Seed Market in Selected Countries for the year 2011*. Nyon, Switzerland: International Seed Federation. [Online]. Available at: http://www.worldseed.org/isf/seed_statistics.html [accessed: 2 Dec. 2012].

Isin, E. 2002. *Being Political: Genealogies of Citizenship*. Minneapolis: University of Minnesota Press.

James, C. 2011. *Global Status of Commercialised Biotech/GM Crops: 2011*. Brief 43. Ithaca: International Service for the Acquisition of Agri-Biotech Applications.

Jasanoff, S. 2005. *Designs on Nature: Science and Democracy in Europe and the United States*. Princeton: Princeton University Press.

Jason, D. 2005. AGM Address, COABC AGM: Seeds for the Future, Sidney, B.C., 25 Feb.

Jones, O. and Cloke, P. 2002. *Tree Cultures: The Place of Trees and Trees in Their Place*. Oxford: Berg Publishers.

Kaplan, A. and Ross, K. 1987. Introduction. *Yale French Studies*, 0(73), 1–4.

Katz, S. 2006. *The Revolution Will Not Be Microwaved*. White River Junction: Chelsea Green Publishing.

Kaskey, J. 2012. Dupont sends in former cops to enforce seed patents. *Bloomberg News*. [Online]. Available at: http://www.bloomberg.com/news/2012-11-28/dupont-sends-in-former-cops-to-enforce-seed-patents-commodities. html [accessed: 29 Nov. 2012].

Kerkvliet, B. 2009. Everyday politics in peasant societies (and ours). *Journal of Peasant Studies*, 36(1), 227–43.

Kloppenburg, J. Jr. 2004. *First the Seed: The Political Economy of Plant Biotechnology, 1492–2000*, second edition. Madison: University of Wisconsin Press.

———. 2010. Impending dispossession, enabling repossession: biological open source and the recovery of seed sovereignty. *Journal of Agrarian Change*, 10(3), 367–88.

Kneen, B. 1993. *From Land to Mouth: Understanding the Food System*, second edition. Toronto: NC Press Ltd.

———. 1999. *Farmageddon: Food and the Culture of Biotechnology*. Gabriola Island: New Society Publishers.

———. 2006. *The Tyranny of Rights*. Ottawa: Ram's Horn.

Kozub, D. 2012. A seed sector value chain roundtable: part 2. *Seed to Succeed*, Spring, 14–15.

Kuhlmann, A. 2009. Chair's report. *Saskatchewan Flax Grower Newsletter*, Sept. Saskatoon: Saskatchewan Flax Development Commission.

Kuyek, D. 2004a. *Stolen Seeds: The Privatisation of Canada's Agricultural Biodiversity*. Sorrento: The Ram's Horn.

———. 2004b. *Reaping What's Sown: How the Privatization of the Seed System will Shape the Future of Canadian Agriculture*. M.A. thesis, Univeristé du Québec, Montréal.

La Via Campesina. 1996. *Plea for Recognition of Farmers' Rights*, 1996 Intervention to the FAO/CGRFA, Rome, Italy, 10 Dec. [Online]. Available at: http://www.ukabc.org/Via_Camp.html [accessed: 13 Oct. 2012].

———. 2003. *People's Food Sovereignty: WTO Out of Agriculture*. 2 Sept. [Online]. Available at: http://viacampesina.org/en/index.php/main-issues-mainmenu-27/food-sovereignty-and-trade-mainmenu-38/396-peoples-food-sovereignty-wto-out-of-agriculture [accessed: 13 Oct. 2012].

———. 2011. *Bali Seed Declaration: Peasant Seeds: Dignity, Culture and Life: Farmers in Resistance to Defend their Right to Peasant Seeds*. 16 Mar. [Online]. Available at: http://viacampesina.org/en/index.php/main-issues-mainmenu-27/biodiversity-and-genetic-resources-mainmenu-37/1030-peasant-seeds-dignity-culture-and-life-farmers-in-resistance-to-defend-their-right-to-peasant-seeds [accessed: 2 Oct. 2012].

Labrada, H. Rios. 2006. From participatory plant breeding to local innovation in Cuba and Mexico, presentation at *From Seeds of Survival to Seeds of Resilience Gathering*, Addis Ababa, Ethiopia, 31 Oct.–6 Nov.

LaDuke, W. and Scott, L. 2011. Native farmers gather to protect seeds. *The Circle News*. 12 Apr. [Online]. Available at: http://thecirclenews.org/index.php?option=com_content&task=view&id=488&Itemid=75 [accessed: 4 Dec. 2012].

Lammerts van Bueren, E.T., Hulscher, M., Haring, J., Jongerden, Ruivenkamp, G.T.P., van Mansvelt, J.D. and den Nijs, A.M.P. 1999. *Sustainable Organic Plant Breeding: Final Report: A Vision, Choices, Consequences and Steps*. Driebergen: Louis Bolk Instituut.

Latour, B. (as Jim Johnson). 1988. Mixing humans with non-humans: sociology of a door-closer. *Social Problems*, 35(3), 298–310.

Latour, B. 1991. Technology is society made durable, in *A Sociology of Monsters: Essays on Power, Technology and Domination*, edited by J. Law. London: Routledge, 103–31.

———. 2004a. *Politics of Nature: How to Bring the Sciences into Democracy.* Boston: Harvard University Press.

———. 2004b. How to talk about the body? the normative dimension of science studies. *Body and Society*, 10(2–3), 205–29.

———. 2005. *Reassembling the Social: An Introduction to Actor-Network Theory.* Oxford: Oxford University Press.

Latour, B. and Woolgar, S. 1979. *Laboratory Life: The Construction of Scientific Facts.* Princeton: Princeton University Press.

Lave, J. 2009. The practice of learning, in *Contemporary Theories of Learning: Learning Theorists in their Own Words*, edited by K. Illeris. London: Routledge, 200–208.

Lave, J. and Wenger, E. 1991. *Situated Learning: Legitimate Peripheral Participation.* Cambridge: Cambridge University Press.

Law, J. 2002. Objects and space. *Theory, Culture and Society*, 19(5/6), 91–105.

———. 2004. *Enacting Naturecultures: A Note from STS.* Lancaster: Centre for Science Studies, Lancaster University.

Law, J. and Mol, A. 2008. The actor-enacted: Cumbrian sheep in 2001, in: *Material Agency, Toward a Non-Anthropocentric Approach*, edited by L. Malafouris and C. Kanppett. New York: Springer, 57–77.

Law, J. and Urry, J. 2004. Enacting the Social. *Economy and Society*, 33(2), 390–410.

Leahy, S. 2005. Ban endures on Terminator seeds. *InterPress Service*. 11 Feb. [Online]. Available at: http://www.ipsnews.net/2005/02/environment-ban-endures-on-terminator-seeds/ [accessed: 17 Nov. 2012].

Leask, W. 1998. Discussion: Canadian Seed Trade Association, in *Proceedings of the 4th Agricultural and Food Policy Systems Information Workshop*, edited by R.M.A. Loyns, R.D. Knutson and K. Meilke. Texas A&M University and University of Guelph, 93–5.

LeBuanec, B. 2005. *Farm-Saved Seed Survey.* Presentation of the International Seed Federation to the Meeting on Enforcement of Plant Breeders' Rights. UPOV/Enforcement/05/3. 25 Oct.

Lefebvre, H. 1984. *Everyday Life in the Modern World*, trans. S. Rabinovitch. London: Transaction Publishers.

———. 2002. Towards a leftist cultural politics, in *Marxism and the Interpretation of Culture*, edited by C. Nelson and L. Grossberg. Chicago: University of Illinois Press, 75–88.

Leidner, R. 1993. *Fast Food, Fast Talk: Service Work and the Routinization of Everyday Life.* Los Angeles: University of California Press.

Leopold, A. 1949. *A Sand County Almanac and Sketches Here and There.* New York: Oxford University Press.

Levkoe, C. 2006. Learning democracy through food justice movements. *Agriculture and Human Values*, 23(1), 89–98.

Light, A. 2007. Restorative relationships: from artifacts to 'natural' systems, in *Healing Natures, Repairing Relationships*, edited by R. France. Cambridge: Green Frigate Books, 95–116.

Lockie, S. 2002. 'The invisible mouth': mobilizing 'the consumer' in food production-consumption networks. *Sociologia Ruralis*, 42(4), 278–94.

Lorimer, H. 2006. Herding memories of humans and animals. *Environment and Planning D*, 24(4), 497–518.

———. 2010. Forces of nature, forms of life: calibrating ethology and phenomenology, in *Taking Place: Non-Representational Theories and Geography*, edited by B. Anderson and P. Harrison. Farnham: Ashgate, 55–78.

Lorimer, J. 2007. Nonhuman charisma. *Environment and Planning D*, 25, 911–32.

———. 2008. Counting corncrakes: the affective science of the UK corncrake census. *Social Studies of Science*, 38(3), 377–405.

———. 2010. Elephants as companion species: the lively biogeographies of Asian elephant conservation in Sri Lanka. *Transactions of the Institute of British Geographers*, 35, 491–506.

Louwaars, N.P., Dons, H., van Overwalle, G., Raven, H., Arundel, A., Eaton, D. and Nelis, A. 2009. *Breeding Business: The Future of Plant Breeding in Light of Developments in Patent Rights and Plant Breeders' Rights*. CGN Report 2009–14. Wageningen: Centre for Genetic Resources.

Louwaars, N.P., Tripp, R., Eaton, D., Henson-Apollonio, V., Hu, R., Mendoze, M., Muhhuku, E., Pal, S. and Wekundah, J. 2005. *Impacts of Strengthened Intellectual Property Rights Regimes on the Plant Breeding Industry in Developing Countries*. Wageningen: Centre for Genetic Resources. [Online]. Available at: http://documents.plant.wur.nl/cgn/literature/reports/IPR%20in%20breeding%20industry.pdf [accessed: 17 Oct. 2012].

Lulka, D. 2004. Stabilizing the herd: fixing the identity of nonhumans. *Environment and Planning D*, 22, 439–63.

Lupescu, M. 2011. *Canada: Agricultural Biotechnology Annual Report*. CA11039. Global Agricultural Information Network, USDA Foreign Agricultural Office. [Online]. Available at: http://gain.fas.usda.gov/Recent%20GAIN%20Publications/Agricultural%20Biotechnology%20Annual_Ottawa_Canada_07-15-2011.pdf [accessed: May 2012].

MacRae, R. 1999. Not just what, but how: creating agricultural sustainability and food security by changing Canada's agricultural policy making practice. *Agriculture and Human Values*, 16, 187–201.

Magdoff, F., Foster, J.B., Buttel, F.H. 2000. An Overview, in *Hungry for Profit: The Agribusiness Threat to Farmers, Food and the Environment*, edited by F. Magdoff, J.B. Foster and F.H. Buttel. New York: Monthly Review Press, 7–21.

Marder, M. 2012. Plant intentionality and the phenomenological framework of plant intelligence. *Plant Signaling and Behavior*, 7(11), 1–8.

Martin, C. 2000. *A History of Canadian Gardening*. Toronto: McArthur and Company.

Marvier, M. and Van Acker, R.C. 2005. Can crop transgenes be kept on a leash? *Frontiers in Ecology and the Environment*, 3(2), 93–100.

Mascarenhas, M. and Busch, L. 2006. Seeds of change: intellectual property rights, genetically modified soybeans and seed saving in the United States. *Sociologia Ruralis*, 46(2), 122–38.

Mattera, P. 2004. *USDA Inc.: How Agribusiness has Hijacked Regulatory Policy at the US Department of Agriculture*. Washington, DC: Agribusiness Accountability Initiative and Corporate Research Project, Good Jobs First. [Online]. Available at: http://www.nffc.net/Issues/Corporate%20Control/USDA%20INC.pdf [accessed: 12 Nov. 2012].

Mauro, I., MacLaughlan, S., and Van Acker, R. 2009. Farmer knowledge and a priori risk analysis: pre-release evaluation of genetically modified Roundup Ready wheat across the Canadian Prairies. *Environmental Science and Pollution Research*, 16, 689–701.

Mazhar, F. 2007. Rights panel discussion. *Seedling*, Oct, 13–16.

McAfee, K. 2003. Neoliberalism on the molecular scale: economic and genetic reductionism in biotechnology battles. *Geoforum*, 34, 203–19.

———. 2008. Beyond techno-science: transgenic maize in the fight over Mexico's future. *Geoforum*, 39(1), 148–60.

McMichael, P. 2005. Global development and the corporate food regime, in, *New Directions in the Sociology of Global Development*, edited by F.H. Buttel and P. McMichael. Oxford: Elsevier Press, 269–303.

———. 2007. Feeding the world: agriculture, development and ecology, in *Socialist Register 2007: Coming to Terms with Nature*, edited by L. Panitch and C. Leys. New York: Monthly Review Press, 170–94.

———. 2008. Peasants make their own history, but not just as they please... *Journal of Agrarian Change*, 8(2–3), 205–28.

———. 2009. A food regime genealogy. *The Journal of Peasant Studies*, 36(1), 139–69.

Mesthene, E. 1983. Technology and wisdom, in *Philosophy and Technology: Readings in the Philosophical Problem of Technology*, edited by C. Mitcham and R. Mackey, R. New York: The Free Press, 109–29.

Mgbeoji, I. 2006. *Global Biopiracy: Patents, Plants, and Indigenous Knowledge*. Vancouver: UBC Press.

Middleton, J. 2010. Sense and the city: exploring the embodied geographies of urban walking. *Social and Cultural Geography*, 11(6), 575–96.

Miele, M. 2011. The taste of happiness: free-range chicken. *Environment and Planning A*, 43(9), 2076–90.

Mignouna, H., Abang, M. and Asiedu, R. 2004. Harnessing modern biotechnology for tropical tuber crop improvement: yam (Dioscorea spp.) molecular breeding. *African Journal of Biotechnology*, 2(12), 478–85.

Miller, J., Salazar, M., Mascarenhas, M. and Busch, L. 2006. The indivisibility of science, policy & ethics: Starlink corn and the making of standards, in *Agricultural Standards: The Shape of the Global Food and Fiber System*, edited by J Bingen and L. Busch. New York: Springer, 111–24.

Mann, A., Mol, A., Satalkar, P., Savirani, A., Selim, N., Sur, M. and Yates-Doerr, E. 2011. Mixing methods, tasting fingers: notes on an ethnographic experiment. *Journal of Ethnographic Theory*, 1(1), 221–43.

Mol, A. 1999. Ontological politics: a word and some questions, in *Actor Network Theory and After*, edited by J. Law and J. Hassard. Oxford: Blackwell, 74–89.

———. 2002. *The Body Multiple: Ontology in Medical Practice*. Durham: Duke University Press.

———. 2008. I eat an apple: on theorizing subjectivities. *Subjectivity*, 22, 28–37.

Monsanto Canada Inc. v. Schmeiser. 2004. SCC 34. Supreme Court of Canada.

Monsanto Company. 2005a. *Monsanto Completes Acquisition of Seminis*. 23 Mar. [Online]. Available at: http://monsanto.mediaroom.com/index.php?s=27632&item=76579 [accessed: 2 Oct. 2012].

———. 2005b. *Annual Report*. St. Louis: Monsanto Company.

———. 2006a. *Annual Report*. St Louis: Monsanto Company.

———. 2006b. *Technology Use Agreement*. St Louis: Monsanto Company.

———. 2007. *Serving Farmers: Seeds and Genomics*. Accessed at: www.monsanto.com, no longer available.

———. 2008. *Annual Report*. St Louis: Monsanto Company.

———. 2009. *Monsanto Company Invests in Developing New Technologies for Wheat with Acquisition of WestBred Business*. 14 Jul. [Online]. Available at: http://monsanto.mediaroom.com/index.php?s=27632&item=77076 [accessed: 2 Oct. 2012].

———. 2012. *Improving Agriculture, Improving Lives*. [Online]. Available at: http://www.monsanto.com/improvingagriculture/Pages/default.aspx [accessed: 2 Oct. 2012].

Monsanto Company profile, part III. 2009. *Organic Lifestyle Magazine*. 10(Oct-Nov): 46–51. [Online]. Available at: http://www.organiclifestylemagazine.com/issue-10/monsanto.php [accessed: 2 Oct. 2012].

Mooney, P. 2006a. Panel discussion, at Food Fights! Rights, Knowledge, Diversity, Peterborough, Canada, 9 Feb.

———. Executive Director, ETC Group, personal communication, Nov. 2006.

Moore, E. 2002. The new direction of federal agricultural research in Canada: from public good to private gain? *Journal of Canadian Studies*, 37(3), 112–34.

Morris, M.L. and Bellon, M.R. 2004. Participatory plant breeding research: opportunities and challenges for the international crop improvement system. *Euphytica*, 136, 21–35.

Murphy, J. and Levidow, L. 2006. *Governing the Transatlantic Conflict over Agricultural Biotechnology: Contending Coalitions, Trade Liberalisation and Standard Setting*. New York: Routledge.

Myers, J. 2012. A public breeder's perspective on the closure of the genetic commons, in *6ᵗʰ Organic Seed Growers Conference Proceedings*, edited by K. Hubbard, M. Colley and J. Zystro. Port Townsend: Organic Seed Growers' Alliance, 130–32.

Nabhan, G. 2009. *Where Our Food Comes From: Retracing Nikolay Vavilov's Quest to End Hunger*. Washington DC: Island Press.

Navdanya. 2012. *Seed Freedom: A Global Citizens' Report*. New Delhi: Navdanya. [Online]. Available at: http://www.navdanyainternational.it/images/doc/ Seed%20Freedom.pdf [accessed: 2 Dec. 2012].

Nazarea, V. 1998. *Cultural Memory and Biodiversity*. Tucson: University of Arizona Press.

———. 2005. *Heirloom Seeds and their Keepers: Marginality and Memory in the Conservation of Biological Diversity*. Tucson: University of Arizona Press.

NFU (National Farmers' Union). 2000. *National Farmers' Union Policy on Genetically Modified (GM) Foods*. Saskatoon: NFU.

———. 2005 *The Farm Crisis and Corporate Profits*. Saskatoon: NFU. 30 November.

———. 2010. *Grain Companies Exploit Flax Situation to Tighten Vise on Farmer Seed Saving*. Press release 18 Jan. [Online]. Available at: http://www.gene. ch/genet/2010/Jan/msg00097.html [accessed: 4 Nov. 2012].

———. 2012a. *Federal Government is Weak-Minded on CETA Deal*. Press release. 27 Apr.

———. 2012b. *Farmers, the Food Chain and Agriculture Policies in Canada in Relation to the Right to Food*. Submission to the Special Rapporteur on the Right to Food, Mission to Canada. [Online]. Available at: http://nfu. fairtrademedia.com/sites/www.nfu.ca/files/NFU%20Final%20Report%20 to%20Special%20Rapporteur%20on%20the%20Right%20to%20 Food,%20May%202012_0.pdf [accessed: 12 Dec. 2012].

NordGen (Nordic Genetic Resource Center). nd. *Agreement between (Depositor) and the Royal Norweigan Ministry of Agriculture and Food Concerning the Deposit of Seeds in the Svalbard Global Seed Vault*. Alnarp: NordGen. [Online]. Available at: http://www.nordgen.org/sgsv/files/ SGSV_Deposit_Agreement.pdf [accessed: 17 Oct. 2012].

———. 2012a. *Svalbard Global Seed Vault: Depositors and Material*. [Online]. Available at http://www.nordgen.org/sgsv/index.php?page=sgsv_ information_sharing [accessed: 31 Oct. 2012].

———. 2012b. *Svalbard Global Seed Vault: Depositor Institutes*. [Online]. Available at: http://www.nodgen.org/sgsv/index.php?app=data_unit&unit=sgsv_by_depositor [accessed: 31 Oct. 2012].

Norse, E. and Carlton, J. 2003. World-wide buzz about biodiversity. *Conservation Biology*, 17(6), 1475–6.

OAPF (Organic Agriculture Protection Fund). 2008. *Individual Action Not the Way to Go: Organic Farmers*. Press Release. 16 Apr.

OCM (Organisation for Competitive Markets). 2008. *Monsanto Corn Seed Price Hikes a Threat to Agriculture*. 24 Jul. [Online]. Available at: http://www.competitivemarkets.com/monsanto-corn-seed-price-hikes-a-threat-to-agriculture/ [accessed: 11 Oct. 2012].

Office of the Auditor General of Canada. 2004. *Canadian Food Inspection Agency: Regulation of Plants with Novel Traits*. Ottawa: Government of Canada.

Oyama, S. 2000. *Evolution's Eye: A Systems View of the Biology-Culture Divide*. Durham: Duke University Press.

Paarlberg, R. 2008. *Starved for Science: How Biotechnology is Being Kept Out of Africa*. Boston: Harvard University Press.

Palmer, C. 2006. Stewardship: a case study in environmental ethics, in *Environmental Stewardship: Critical Perspectives, Past and Present*, edited by R.J. Berry. New York: Continuum, 63–75.

Patel, R. 2007. Transgressing rights: La Via Campesina's call for food sovereignty. *Feminist Economics*, 13(1), 87–93.

———, ed. 2009. Special section: food sovereignty. *Journal of Peasant Studies*, 36(3), 663–706.

———. 2010. What does food sovereignty look like? in *Food Sovereignty: Reconnecting Food, Nature, and Community*, edited by H. Wittman, A.A. Demarais and N. Wiebe. Winnipeg: Fernwood Publishing, 186–96.

Patent Act, 1985, cP-4. Ottawa, Government of Canada.

Patent Appeal Board of Canada.1982. *Re Application for Patent of Abitibi Co.*, 62 CPR, 2d.

Pearce, F. 2005. Return to Eden. *New Scientist*, 185(2483), 35–7.

Pendleton, C.N. 2004. The peculiar case of 'Terminator' technology. *Biotechnology Law Report*, 23, 1–29.

Pfeiffer, J.M., Rice, K.J., Dun, S. and Mulawarman, B. 2006. Biocultural diversity in traditional rice-based agroecosystems. *Environment, Development, Sustainability*, 8, 609–25.

PGRC (Plant Gene Resources Canada). 2010. Major Holdings. Ottawa: PGRC. [Online]. Available at: http://pgrc3.agr.gc.ca/holdings-stocks_e.html [accessed: 22 Nov. 2012].

———. nd. Operational Sequence for Handling Seed Samples in PGRC. Saskatoon: Plant Gene Resources Canada.

Phillips, C. 2005. Cultivating practices: saving seed as green citizenship? *Environments*, 33(3), 37–49.

———. 2008. Canada's evolving seed regime: relations of industry, state and seed savers. *Environments*, 36(1), 5–18.

———. forthcoming. Living without fruit flies: biosecuring horticulture and its markets. *Environment and Planning A*.

Phillipson, M. 2005. Giving away the farm? the rights and obligations of biotechnology multinationals: Canadian developments. *The King's College Law Journal*, 16(2), 362–72.

Pioneer Hi-Bred Ltd. v. Canada (Commissioner of Patents). 1989. 1SCR 1623. Supreme Court of Canada.

Pistorius, R. 1997. *Scientists, Plants and Politics: A History of the Plant Genetic Resources Movement*. Rome: International Plant Genetic Resources Institute.

Pistorius, R. and van Wijk, J. 1999. *The Exploitation of Plant Genetic Information: Political Strategies in Crop Development*. Oxon: CABI.

Plant Breeders' Rights Act, 1990, c20. Ottawa: Government of Canada.

Plumwood, V. 2001. Nature as agency and the prospects for a progressive naturalism. *Capitalism Nature Socialism*, 12(4), 3–32.

———. 2002a. Prey to a crocodile. *Aisling Magazine*, 30. [Online]. Available at: http://www.aislingmagazine.com/aislingmagazine/articles/TAM30/Contents.html [accessed: 22 Oct. 2012].

———. 2002b. *Environmental Culture: The Ecological Crisis of Reason*. London: Routledge.

———. 2007. A review of Deborah Bird Rose's 'Reports from a Wild Country: ethics of decolonisation'. *Australian Humanities Review*, 42, 1–4.

———. 2009. Nature in the Active Voice. *Australian Humanities Review*, 46, 113–29.

Podoll, T. 2012. Rebuilding diversity and resilience in a concentrated seed sector, in *6th Organic Seed Growers Conference Proceedings*, edited by K. Hubbard, M. Colley and J. Zystro. Port Townsend: Organic Seed Growers' Alliance, 133–9.

Pollan, M. 1991. *Second Nature: A Gardener's Education*. New York: Dell Publishing.

Potts, A. and Haraway, D. 2010. Kiwi chicken advocate talks with Californian dog companion. *Feminism & Psychology*, 20(3), 318–36.

Power, E. 2005. Human-nature relations in suburban gardens. *Australian Geographer*, 36(1), 39–53.

Power, T.M. 2000. Trapped in consumption: modern social structure and the entrenchment of the device, in *Technology and the Good Life?*, edited by E. Higgs, A. Light and D. Strong. Chicago: University of Chicago Press, 271–93.

Pratt, S. 2010a. Certified flax seed: only idea splits growers. *Western Producer*. 28 Jan.

———. 2010b. NDP MP's Bill worries canola industry. *Western Producer*. 21 Jan.

———. 2010c. Dow warns against GMO Bill. *Western Producer*. 30 Sept.

Probyn, E. 2011a. Swimming with tuna: human-ocean entanglements. *Australian Humanities Review*, 51, 237–86.

———. 2011b. Eating roo: of things that become food. *New Formations*, 74, 33–45.

Prudham, S. 2007. The fictions of autonomous invention: accumulation by dispossession, commodification and life patents in Canada. *Antipode*, 39(3), 406–29.

Quist, D. and Chapela, I.H. 2001. Transgenic DNA introgressed into traditional maize landraces in Oaxaca, Mexico. *Nature*, 414, 541–3.

Raghavan, C. 2000. Neem patent revoked by European Patent Office. *South-North Development Monitor*.

Rajotte, T. 2008. The negotiations web: complex connections, in *The Future Control of Food*, edited by G. Tansey and T. Rajotte. London: Earthscan, 141–67.

Rao, N.K, Hanson, J., Dulloo, M.E., Ghosh, K., Nowell, D. and Larinde, M. 2006. *Manual of Seed Handling in Genebanks*. Handbook 8. Rome: Biodiversity International.

Reckwitz, A. 2002. Toward a theory of social practices: a development in culturalist theorizing. *European Journal of Social Theory*, 5(2), 243–63.

Reid, I.R. and Mosseler, A. 1995. *Canada: Country Report to the FAO International Technical Conference on Plant Genetic Resources*. Ottawa: Government of Canada.

Reznik, S. and Vavilov, Y. 1997. The Russian scientist Nicolay Vavilov, in *Five Continents*, by N.I. Vavilov, edited by L.E. Rodin, trans. D. Love. Rome: International Plant Genetic Resource Institute.

Richards, K. Manager, Plant Genetic Resources Canada, personal communication, May 2007.

Robbins, P. 2007. *Lawn People: How Grasses, Weeds, and Chemcials Make Us Who We Are*. Philadelphia: Temple University Press.

Robinson, D. 2010. *Confronting Biopiracy: Challenges, Cases and International Debates*. London: Earthscan.

Roe, E.J. 2006a. Material connectivity, the immaterial and the aesthetic of eating practices: an argument for how genetically modified foodstuff becomes inedible. *Environment and Planning A*, 38(3), 465–81.

———. 2006b. Things becoming food and the embodied, material practices of an organic food consumer. *Sociologia Ruralis*, 46(2), 104–21.

Rose, G. 1993. *Feminism and Geography: The Limits of Geographical Knowledge*. Minneapolis: University of Minnesota Press.

Royal Society of Canada. 2001. *Elements of Precaution: Recommendations for the Regulation of Food Biotechnology in Canada.* Ottawa: Royal Society of Canada.

Sackville-Hamilton, N.R. and Chrolton, K.H. 1997. *Regeneration of Accessions in Seed Collections: A Decision Guide.* Handbook 5. Rome: International Plant Genetic Resource Institute.

Scarelli, N., Tostain, S., Vigouroux, Y., Agbangla, C., Daïnou, O. and Pham, J.L. 2006. Farmers' use of wild relative and sexual reproduction in a vegetatively propagated crop: the case of yam in Benin. *Molecular Ecology*, 15, 2421–31.

Schatzki, T. 1996. *Social Practices: A Wittgenstenian Approach to Human Activity and the Social.* Cambridge: Cambridge University Press.

Schmeiser, P. 2005. Public presentation and discussion, Toronto, Canada, 13 May.

Schmidt, S. 2012. Canada ready to unveil plan to ease trade of genetically modified foods. *Postmedia news.* 15 Aug. [Online]. Available at: http://o. canada.com/2012/08/15/h2canada-ready-to-unveil-plan-to-ease-trade-of-genetically-modified-foods/ [accessed: 17 Oct. 2012].

Schneekloth, L. 2001. Plants: the ultimate alien. *Extrapolation*, 42(3), 246–54.

———. 2002. Alien kin: humans and the forest, *Organdi Quarterly*, 5.

Scholz, A. 2004. Merchants of diversity: scientists as traffickers of plants and institutions, in *Earthly Politics: Local and Global in Environmental Governance*, edited by S. Jasanoff and M.L. Martello. Cambridge: MIT Press, 217–38.

Scott, J. 1985. *Weapons of the Weak: Everyday Forms of Peasant Resistance.* London: Yale University Press.

SeedQuest. 2005. *CWB Releases Focus Group Excerpts, News Section.* Nov. [Online]. Availabe at: http://www.seedquest.com/News/from/ northamerica/newsfromcanada/2005.htm [accessed: 15 Nov. 2012].

Seeds Act, 1985, cS-8. Ottawa: Government of Canada.

Seeds Regulations, 2012, CRC, C1400. Ottawa: Government of Canada.

SFDC (Saskatchewan Flax Development Commission). 2009. *Market Support Program.* Nov.

Sharratt, L. Coordinator, Canadian Biotechnology Action Network, personal communication, Aug. 2007.

Shirtliffe, S. Department of Plant Sciences, University of Saskatchewan, personal communication, Aug. 2007.

Shiva, V. 1993. *Monocultures of the Mind: Perspectives on Biodiversity and Biotechnology.* London: Zed Books.

———. 1995. Epilogue, in *Biopolitics: A Feminist and Ecological Reader on Biotechnology*, edited by V. Shiva and L. Moser. London: Zed Books, 267–84.

———. 1997. *Biopiracy: The Plunder of Nature and Knowledge.* Toronto: Between the Lines.

———. 2000. *Stolen Harvest, the Hijacking of the Global Food Supply.* Cambridge: South End Press.

Shove, E. and Pantzar, M. 2005. Consumers, producers and practices understanding the invention and reinvention of Nordic walking. *Journal of Consumer Culture*, 5(1), 43–64.

Shove, E., Trentmann, F. and Wilk, R. 2009. *Time, Consumption and Everyday Life.* Oxford: Berg.

Slinkard, A., Knott, E. and Douglas, R., eds. 1995. *Harvest of Gold: The History of Field Crop Breeding in Canada.* Saskatoon: University of Saskatchewan Extension Press.

Smale, M. and Day-Rubenstein, K. 2002. The demand for crop genetic resources: international use of the US National Plant Germplasm System. *World Development*, 30(9), 1639–55.

Spinney, J. 2006. A place of sense: a kinaesthetic ethnography of cyclists on Mont Ventoux. *Environment and Planning D*, 24(5), 709–32.

SSCAF (Standing Senate Committee on Agriculture and Forestry). 2004. Proceedings, Issue 4, Evidence 9, Dec. First Session, 38th Parliament.

SSE (Seed Savers Exchange). 2004. *Garden Seed Inventory*, sixth edition. Decorah: Seed Savers Exchange.

Staddon, C. 2009. Towards a critical political ecology of human-forest interactions: collecting herbs and mushrooms in a Bulgarian locality. *Transactions of the Institute of British Geographers*, 34(2), 161–76.

Statistics Canada. nd. *Estimated areas, yield, production and average farm price of principal field crops, in metric tonnes, annual.* CANSIM. [Online]. Available at: http://www.statcan.gc.ca [accessed: 17 Oct. 2012].

Stengers, I. 1997. *Power and Invention: Situating Science.* Minneapolis: University of Minnesota Press.

———. 2000. God's heart and the stuff of life. *Pli*, 9, 86–118.

———. 2010. Including nonhumans in political theory: opening Pandora's box?, in *Political Matter: Technoscience, Semocracy, and Public Life*, edited by B. Braun and S. Whatmore. Minneapolis: University of Minnesota Press.

Strathern, M. 2001. The patent and the Malanggan. *Theory, Culture, and Society*, 18(4), 1–26.

———. 2004. *Partial Connections.* Oxford: Rowman and Littlefield Publishers.

Stein, A. and Rodríguez-Cerezo, E. 2010. Low-level presence of new GM crops. *AgBioForum*, 13(2), 173–82.

Symko, S. 1999. *From a Single Seed: Tracing the Marquis Wheat Success Story in Canada to its Roots in the Ukraine.* Ottawa: Agriculture and Agri-Food Canada.

Syngenta Foundation. 2012. *Syngenta Foundation.* [Online]. Available at: http://www.syngentafoundation.org [accessed: 17 Oct. 2012].

Takacs, D. 1996. *The Idea of Biodiversity: Philosophies of Paradise*. Baltimore: John Hopkins University Press.

Tansey, G. 2002. Food for thought: intellectual property rights, food, and biodiversity. *Harvard International Review*, Spring, 54–9.

Tansey, G. and Rajotte, T., eds. 2008. *The Future Control of Food: A Guide to International Negotiations and Rules on Intellectual Property, Biodiversity and Food Security*. London: Earthscan.

ten Kate, K. and Laird, S.A. 1999. *The Commercial Use of Biodiversity: Access to Genetic Resources and Benefit-Sharing*. London: Earthscan.

Thompson, P.B. 1994. *The Spirit of the Soil: Agriculture and Environmental Ethics*. London: Routledge.

Thrift, N. 2000. Still life in the nearly present time: the object of nature. *Body and Society*, 6(3–4), 34–57.

Thrupp, L.A. 1998. *Cultivating Diversity: Agrobiodiversity and Food Security*. Washington, D.C.: World Resources Institute.

———. 2000. Linking agricultural biodiversity and food security: the valuable role of agrobiodiversity for sustainable agriculture. *International Affairs*, 76(2), 265–81.

Tickell, A. and Peck, J. 2003. Making global rules: globalization or neoliberalization?, in *Remaking the Global Economy: Economic-Geographical Perspectives*, edited by J. Peck and H.W. Yeung. London: Sage Publications, 163–81.

Tsing, A. 2005. *Friction: An Ethnography of Global Connection*. Princeton: Princeton University Press.

———. 2012. Unruly edges: mushrooms as companion species. *Environmental Humanities*, 1, 141–54.

UNEP (United Nations' Environment Programme). 2003. *Report of the Ad Hoc Technical Expert Group Meeting on the Potential Impacts of Genetic Use Restriction Technologies on Smallholder Farmers, Indigenous and Local Communities and Farmers' Rights*. Subsidiary Body on Scientific, Technical and Technological Advice, 9th meeting, 10–14 Nov.

———. 2005a. *Compilation of Submissions on Potential Socio-Economic Impacts of Genetic Use Restriction Technologies (GURTs) on Indigenous and Local Communities*. Ad Hoc Open-Ended Inter-Sessional Working Group on Article 8(j) and Related Provisions of the Convention on Biological Diversity. UNEP/CBD/WG8J/4/INF/6.

———. 2005b. *Submission to the Convention on Biological Diversity on Advice on the Report of the Ad Hoc Technical Expert Group on Genetic Use Restriction Technologies*. UNEP/CBD/WG81/4/INF/17.

UPOV. 2005. *UPOV Report on the Impact of Plant Variety Protection*. Geneva: International Union for the Protection of New Varieties of Plants.

USDA (United States Department of Agriculture). 2007. Report on the Status of Sunflower Germplasm in the US. National Plant Germplasm System 1-2007. Washington, DC: USDA.

Van Acker, R.C. 2005. *Co-Existence of GM and Non-GM Crops in Canada: Current Status and Future Direction*. Paper presented at Second International Conference on the Co-Existence between GM and Non-GM Supply Chains, Montpellier, France, 14–15 Nov.

Van Acker, R.C., Szumgalski, A. and Friesen, L. 2007. The potential benefits, risks, costs of genetic use restriction technologies. *Canadian Journal of Plant Science*, 87(4), 753–62.

Van Dooren, T. 2007. Terminated seed: death, proprietary kinship and the production of (bio)wealth. *Science as Culture*, 16(1), 71–94.

———. 2009. Banking seed: use and value in the conservation of agricultural diversity. *Science as Culture*, 18(4), 373–95.

———. 2012. Wild seed, domesticated seed: companion species and the emergence of agriculture. *PAN: Philosophy Activism Nature*, 9, 22–8.

Varela, F. 1999. *Ethical Know-How: Action, Wisdom, and Cognition*. Stanford: Stanford University Press.

Vellvé, R. 1992. *Saving the Seed: Genetic Diversity and European Agriculture*. London: Earthscan.

Vernooy, R. 2003. *Seeds that Give: Participatory Plant Breeding*. Ottawa: International Development Research Centre.

Vidal, J. 2010. Why is the Gates Foundation investing in GM giant Monsanto? *The Guardian*. 29 Sept. [Online]. Available at: http://www.guardian.co.uk/ global-development/poverty-matters/2010/sep/29/gates-foundation-gm-monsanto [accessed: 17 Oct. 2012].

Visser, B., Van der Meer, I., Louwaars, N., Beekwilder, J. and Eaton, D. 2001. *Potential Impacts of Genetic Use Restriction Technologies (GURTs) on Agrobiodiversity and Agricultural Production Systems*. Background Paper 5. Rome: FAO.

Warick, J. 2001. GM flaxseed yanked off Canadian market – rounded up, crushed. *StarPhoenix*. 23 Jun.

Warwick, H. and Meziani, G. 2002. *Seeds of Doubt: North American Farmers' Experiences of GM Crops*. Bristol: Soil Association.

Weis, T. 2007. *The Global Food Economy: The Battle for the Future of Farming*. New York: Zed Books.

Wells, S. 2005. Keynote Address. Presentation at *Canadian Organic Association of British Columbia Annual General Meeting: Seeds for the Future*, Sidney, Canada, 25–7 Feb.

Welsh, J. and MacRae, R. 1998. Food citizenship and community food security: lessons from Toronto, Canada. *Canadian Journal of Development Studies*, 19, 237–55.

Wertz, Spencer K. 2005. Maize: the native North American's legacy of cultural diversity and biodiversity. *Journal of Agricultural and Environmental Ethics*, 18, 131–56.

Whatmore, S. 2002. *Hybrid Geographies: Natures, Cultures, Spaces*. London: Sage.

———. 2006. Materialist returns: practising cultural geography in and for a more-than-human world. *Cultural Geographies*, 13, 600–609.

Whatmore, S. and Thorne, L. 1997. Nourishing networks: alternative geographies of food, in *Globalising Food: Agrarian Questions and Global Restructuring*, edited by D. Goodman and M.J. Watts. London: Routledge, 287–304.

Whatmore, S., Stassart, P. and Renting, H. 2003. Guest editorial: what's alternative about alternative food networks? *Environment and Planning A*, 35, 389–91.

Whealy, D. 2011. *Gathering: Memoir of a Seed Saver*. New York: Chelsea Green Publishing.

Whealy, K. 2010. Svalbard Doomsday Vault: Biopiracy by UN Treaty. Speech given at *Prairie Festival*, The Land Institute, Salina, Kansas. 26 Sept. [Online]. Available at: http://www.centerforfoodsafety.org/wp-content/uploads/2010/11/Land-Inst-Svalbard-portion.pdf [accessed: 17 Nov. 2012].

Wildfong, B. Executive Director, Seeds of Diversity Canada, personal communication, Jun. 2005.

Wilkins, J. 2005. Eating right here: moving from consumer to food citizen. *Agriculture and Human Values*, 22(3), 269–73.

Wilson, B. 2010. Scientists reject market acceptance as GM approval factor. *Western Producer*. 14 Oct. 2010.

Wilson, E.O., ed. 1988. *BioDiversity*. Washington, D.C.: National Academy Press.

Winner, L. 1986. *The Whale and the Reactor: A Search for Limits in an Age of High Technology*. Chicago: University of Chicago Press.

Winson, A. 1993. *The Intimate Commodity: Food and the Development of the Agro-Food Complex in Canada*. Toronto: Garamond Press.

Winter, M. 2004. Geographies of food: agro-food geographies: farming, food and politics. *Progress in Human Geography*, 28(5), 664–70.

Wittman, H., Desmarais, A., Wiebbe, N., eds. 2010a. *Food Sovereignty: Reconnecting Food, Nature and Community*. Winnipeg: Fernwood.

———. 2010b. *Food Sovereignty in Canada: Creating Just and Sustainable Food Systems*. Winnipeg: Fernwood.

Wolch, J. and Emel, J., eds. 1998. *Animal Geographies: Place, Politics, and Identity in the Nature-Culture Borderlands*. London: Verso Press.

Wong, T. and Dutfield, G. 2011. *Intellectual Property and Human Development: current trends and future scenarios*. Cambridge: Cambridge University Press.

Woodhead, E. 1998. *Early Canadian Gardening: An 1827 Nursery Catalogue.* Montréal: McGill-Queens University Press.

Worede, M. International Scientific Advisor, Seeds of Survival, USC Canada, personal communication, Jul. 2006.

Worede, M., Teshome, A. and Tesemma, T. 2000. Participatory approaches linking farmers' access to genebanks: Ethiopia, in *Participatory Approaches to the Conservation and Use of Plant Genetic Resources*, edited by E. Friis-Hansen and B. Sthapit. Rome: International Plant Genetic Resources Institute, 56–61.

Worldwatch Institute. 2012. *Genetically Modified Crops Only a Fraction of Global Crop Production.* [Online]. Available at: http://www.worldwatch.org/node/5950. [accessed: 7 Dec. 2012].

Wright, B.D. and Pardey, P.G. 2006. Changing intellectual property regimes: implications for developing country agriculture. *International Journal of Technology and Globalisation*, 2(1), 93–114.

Wylie, J. 2002. An essay on ascending Glastonbury tor. *Geoforum*, 33(4), 441–54.

———. 2005. A single day's walking: narrating self and landscape on the South West Coast Path. *Transactions of the Institute of British Geographers*, 30(2), 234–47.

Yusoff, K. 2009. Excess, catastrophe, and climate change. *Environment and Planning D*, 27(6), 1010–29.

———. 2010. Biopolitical economies and the political aesthetics of climate change. *Theory, Culture and Society*, 27(2–3), 73–99.

Zerbe, N. 2004. Feeding the famine? American food aid and the GMO debate in Southern Africa. *Food Policy*, 29, 593–608.

———. 2007. Contesting privatization: NGOs and farmers' rights in the African model law. *Global Environmental Politics*, 7(1), 97–119.

Index